Biology

The Ultimate Self-Teaching Guide -

Introduction to the Wonderful World of Biology

2nd Edition

by

Bobbi Leigh Templeton

Contents

Introduction:

Biology can be defined as the study of all the processes, structures and growth that contribute to life. This includes plant life, animal life and especially human life. There are so many processes involved in maintaining life. Some of these start at the cellular level, but they continue through how organs interact with each other.

But the extent of biology goes beyond just the bodies of our plants, animals and ourselves. The earth itself is a model of biology at work. The bioprocesses that keep the planet itself functioning are also play a significant part of biology. Imagine for a moment the earth's natural recycling function ceased and nothing was being properly broken down. The biological function that allows for nutrients to be returned to the earth would not be working. Thus, the earth would be unable to continue to produce the necessary plant life, which serves as food for not only humans, but animals as well.

The interactions between plants, animals, bugs and humans are one of the most significant parts of biology itself. Without the give and take between the various species, many of the world's greatest animals would simply cease to exist. The question is how does biology work at the molecular or cellular level? This is the first step in understanding other interactions that together make up the biology of our world.

Throughout this book, we will discuss the various levels of biology and some of the interactions on each level. While this is not an exhaustive list, it definitely provides a starting point for you to dive into the wonderful world of nature via its biology.

Chapter One:
At the Cellular Level: Molecules

The most basic functions of life itself is happen on a level that we cannot see with the naked eye. Cellular functions are typically visible only because of the magnifying power of the microscope. Yet without the functions, life itself would simply become non-existent.

What are a few of these necessary functions? One of the first and most important is the cycle of a typical cell. These are the building blocks of everything on the planet, so understanding how the develop, their life cycle and finally how they are recycled themselves is a necessary first step for anyone starting out in the study of life or biology.

The first step is to understand what makes up a cell. It has a variety of parts that each have their own specific function or duties within the cell. So what are these parts of a cell?

The primary cell of all life is called a eukaryotic, which has very defined compartments within its borders. A plasma membrane and a cell wall (which is primarily found in plants versus animals) provide the natural border of a cell. These border is semi-permeable, meaning that it allows the entrance and exit of various materials. Consider this border the gatekeeper of the cell, regulating the cell's environment and maintaining its electrical potential. Cytoplasm is the material that helps the cell keep its shape and also provide the material that everything within the cell travels through. Cytoskeleton is another part of the structure that anchors the compartments in place, as well as assisting in keeping the cell's necessary shape. Call it the public transportation of any cell.

However, a public transportation system needs a control center, which is what the DNA of a cell functions as. The DNA or chromosomes of a cell are housed within its nucleus. This provides the cell a set of blueprints. These blueprints or plans not only tell the cell what its function will be within the body, but along with RNA also gives the necessary directions for that cell to divide and create copies of itself. Truly, the DNA, RNA and the nucleus are the brains of the operations of the cell. We will discuss DNA in more detail later.

Imagine a muscle cell. The DNA housed in the nucleus will tell that cell how to perform as part of a muscle, versus giving it the direction of behaving as if it were a cell within the stomach. Without these instructions, the cell would be unable to properly complete its various tasks. So what are some of the tasks a cell completes and how does it relate to the compartments that make up the cell itself?

Compartments of Cells

The compartments of a cell are typically called organelles. These organelles are the source of metabolic functions that occur in each cell. Mitochondria and chloroplasts are the powerhouse of the cells, providing energy. Chloroplasts are found particularly in plants, while mitochondria are found within the cells of animals and humans.

ER or endoplasmic reticulum is the transportation network for particles targeted to specific destinations or to join with specific proteins. Golgi apparatus is similar to the wrapping department at a store during the holidays. They package various proteins and lipids after being synthesized by the cell. Inside a cell, digestion is also a part of daily operations. Lysosomes provide the acids for digestion, which means digesting old cell parts, as well as viruses and bacteria when they invade a cell. In many ways, these provide a cleaning and security service all in one at the cellular level.

Vacuoles are the waste storage facility, until the cell is able to send the waste out of the cell itself. For plants, vacuoles are also serve a water storage function at the cellular level. When plants stalks are standing upright, part of that structure definition comes from these water containers. Finally, there are ribosomes. These are complex combinations of both RNA and various proteins. Consider this a factory where various proteins are synthesized into amino acids under the direction of RNA.

Now that we have a basic idea of the organelles with a cell, let's talk about the most important part of our cells, the DNA.

Chapter Two:
Structure of DNA

DNA is a complex structure, made up of nucleic acids. These are information carriers, along with RNA the other nucleic acid. These two combine to create the guidance system for cells. As a result, they provide the knowledge to create specific proteins that define specific biological traits for both individuals, plants and humans. This information gets passed onto each generation of cells, thus making sure our eyes are a specific color or the shape of our nose.

While models of DNA appear very complicated, these are actually get various patterns of four nucleotides. It is an alphabet of sorts, but with only four specific letters. Each of these letters pairs with another specific letter. As they join together, they bind hydrogen molecules with sugar molecules and phosphate molecules. The result is a twisting ladder of sorts. In order to fit into the cell's nucleus, the DNA continues to twist around itself, creating a circular chromosome.

For example, humans have 46 chromosomes. When a specific non-sex human cell reproduces, it receives two copies of each set of chromosomes. But a sex cell receives only one copy of each set of chromosomes. The reason is that when the sperm and egg combine, they will then together have the required two sets of chromosomes. If they had multiple copies, it would be unable to bind property to create a new human.

Before the DNA can pass on the information to a new cell, it needs to create a duplicate of itself. Essentially, it unzips, makes a copy by binding new nucleotides to each side of the unzipped DNA. The process begins when a specific enzyme nicks the double helix, causing them to begin to separate.

Small proteins bind to the keep the DNA unzipped, while other enzymes begin the process of bind new nucleotides to both sides. Another enzyme does the proofreading, making sure that both copies are accurate. Eventually, the DNA is sealed and curls back up.

Once the copy of DNA is complete, the cell begins the process of duplicating everything else, including the organelles. Then the nucleus splits and two new cells are born.

So what does DNA carry the instructions for? Basically, the DNA provides the information needed for the body to create a variety of proteins, specific biological characteristics and even instructions on how to handle poisonous substances, like snake venom. Other information includes how to handle hormones, create storage facilities within the cells and especially protective proteins that deal with a variety of invaders.

Three nucleotides create a word and each of them defines with amino acids are part of each protein. There are 64 potential codon patterns, but only 20 amino acids, so some replication does occur within the genetic code. Still, by use of these patterns, we can look in the mirror and see all the variety and detail these patterns can provide. Then by alternating the various codon, other patterns can emerge. Depending on the protein, it can take anywhere from 100 to 1,000 different codons to create it. The amount of detail and work to create the smallest detail in a plant, animal or human is incredible.

As in any building process, there is a starting and stopping point. Thus, when a list of codons comes together to create a protein, it also has a starting point and an enzyme that tells the codon list it is complete.

So how does a protein actually get produced? First a working copy of DNA is produced, called the mRNA. It then heads to a specific area of the cell to begin production. Ribosomes are essentially the assembly line workers and they use the detail from the mRNA to create the various proteins. The end result is that thousands of proteins are being produced constantly by the body.

Different cells also divide at different rates. Some divide for a specific number of times and then they stop dividing and simply die off. Other are constantly replicating themselves. Still others can be forced to divide based on injury or potential need. The body is constantly monitoring the production of both proteins and cells to keep the body's balance in check.

Now that we have a basic understanding of what is occurring at the molecular level, let's move up to the organs, which are made of large groups of these cells and proteins.

Chapter Three:
Understanding the Functions of Organs

Everyone comes with multiple organs. From skin to livers, even our eyes, all these organs are necessary parts of a functioning human being. Animals also have organs, while plants maintain a more cellular perspective, not having what we would necessarily classify as organs. Since there are so many organs, we are going to concentrate on just one internal and one external. Thus we can learn how they operate and some of their basic functions. Yet many of the attributes of these organs can be applicable to many other organs as well.

The internal organ for our discussion is the liver. This organ is typically located below the diaphragm in most humans, or the upper right side of their abdomen. One of the primary functions of the liver is to assist in the detoxification of the body. The liver has the unique distinction of being the biggest internal organ, as opposed to our skin that is external, and also the biggest gland within the body. It releases hormones that assist in digestion, as well as breaking down insulin and other necessary hormones as part of the body's typical functions. Ducts within the liver allow it to deliver the end results of these various processes, including the production of bile.

Within the liver is the production of proteins, as well as in the metabolism of a variety of molecules. For example, the liver is pivotal in carbohydrate metabolism, turning glucose into glycogen. It also is responsible for synthesizing glucose when the body needs it, which is known as gluconeogenesis. In order to do so, the liver needs to breakdown fat to get at the glycerol that is stored there. Essentially, the liver is a large factory, constantly bringing in and sending out materials according to the orders from the body. While it preforms multiple jobs, this is just one example of how the liver balances to the demands of the body, constantly making adjustments.

Additionally, the liver itself can be regrown. As a result, this is one of the few organs that can be transplanted from a living donor several times. Doctors are constantly studying this unique organ to better understand all the jobs it undertakes during a typical human lifecycle.

How can this translate to other organs? Similar to the liver, other organs perform specific duties based on the needs of the body. The brain itself is the command center of the body. It uses specific hormones as messengers to the rest of the organs, telling them when more energy is needed, when it is time to start general or specific repairs and even when to signal that we are hungry or thirsty. No matter what is going on in the body, there is an organ being called into play.

The largest of our external organs is the skin. This organ provides our bodies the first line of defense against potential invading viruses and bacteria. As with the cells, the skin is a gatekeeper. Still, skin can be damaged. Wounds are essentially breaches of the gatekeeper's defenses. Scabs form from dried blood and other elements to create a patch in the skin's defenses while on the cellular level, the skin goes to work repairing itself. The skin is also in a constant state of regeneration. The oldest cells are found on the top of the skin. As these are scraped off during daily activities, newer skin cells emerge. The skin itself pushed older skin cells off as the newer skins cells are created below. It is similar to cars on a roller coaster, just before it crests the top hill. Each car goes over the top, pushed in some respects by the cars behind.

Both of these organs produce hormones, provide specific functions for the body and are also involved in cellular reproduction as we described in Chapter One. Yet there are clearly specific functions that are only provided by a distinct organ. Examples include the eyes, stomach and kidneys. Each of these completes very specific functions. The eyes are essentially a camera recording the world for the brain. The stomach is a critical part of the digestion process, rending food into a broken down form that allows the body to use the nutrients at the cellular level. Finally, the kidneys work in harmony with the liver to keep the body clear of toxics via a continuous process of filtering the blood.

Yet none of these organs can function without the others within the body. Essentially, the body is one working organism, a true sum of its parts. Biology is a constant study of how these parts work together. As we have seen, starting at the cellular level, all the processes of life require a team effort. Nothing functions without another protein, organ or even another organism. As we will discuss in later chapters, the interdependency of organisms with each other and their environment can have drastic consequences when the environment is altered in any way.

Additionally, organisms themselves follow a specific pattern known as a lifecycle. So what is the typical lifecycle of most organisms, including plants and animals? The following chapters will look at the lifecycle of an organism, but also the various messenger systems available in the body.

Chapter Four:
Getting to Understand How Hormones Function

Many of the daily functions that occur in our bodies are triggered due to hormones. Every organ has glands that contribute to the body's stock of these special triggers. Within our brains are several glands whose only jobs are to create regulating hormones. So how do these hormones actually work within our bodies?

The framework of hormones starts on the molecular level. Our cells not only produce additional cells and other building blocks, but they are the forefront of hormone products. The fundamental job of hormones is to serve as one of two major messenger systems within the body. While many creatures also have a nervous system, the hormones can travel down into the body at the cellular level. It is this ability to provide information at the cellular level that make it a critical part of the body's functioning and overall homeostasis maintenance.

Most hormones are secreted directly into the blood from glands that are part of the endocrine system. Endocrine actually means secreting internally, which aptly describes the way the glands and hormones work together. When an animal or human is exposed to chemicals that interfere with the hormones ability to deliver their messages, these are called endocrine disruptors. These disruptions can alter growth patterns of the body, as well as the delivery of other necessary messages. When this occurs, it has been known to cause other diseases and medical issues, both short and long term. The effects can be intense and individuals may struggle for years as a result.

As a result, endocrine disruptors and their role in various diseases continues to be a major source of study within both the biological and medical fields. Let's look at one particular set of glands that might be most familiar as an example of how these glands and hormones work together to regulate myriads of functions within the body.

The ovaries and testes are the most familiar endocrine glands, as they are distinctly tied to the reproduction aspects of most creatures, especially humans. Each of these glands assist in producing the eggs and sperm for reproduction, but they also serve as the sources of sex hormones that regulate puberty and other aspects of the physical maturing of an individual. This includes the sex drive, as well as the development and maturing of reproduction organs.

These hormones do more than just signal for hair growth in unusual places or the budding of breasts in females. The amount of these hormones can also determine secondary sex characteristics, including muscle mass or how much facial hair you will grow. In addition, they assist in the production of sperm, regulating the menstruation cycle and even the various factors related to pregnancy. Babies are protected by means of these hormonal adjustments and allowed to grow normally as a result.

Therefore, when these glands suffer damage or are not properly formed, the amount of hormone necessary to complete these functions is not available. Hence the reason for tests that check your hormone levels when you have various medical issues. Synthetic hormones have produced by the pharmaceutical companies to assist individuals who struggle with below average hormone levels.

Other types of endocrine glands include the pancreatic islets, adrenal and thyroid. Two diseases directly related to the thyroid are hyperthyroidism and hypothyroidism. When the thyroid is in hyper mode, it makes too much of its hormone. Hence, this contributes to multiple health problems for those who suffer with this problem. Since the thyroid controls metabolism, those with the hyperthyroidism will struggle to gain weight, have an increased heartbeat, struggle to maintain temperature and have fluid moods that range from moody to nervous. However, with treatment, these symptoms can be controlled to help the thyroid slow down.

Hypothyroidism is the opposite problem. The thyroid gland does not make enough hormone, resulting in the body gaining excess weight and being tired all the time. This also is treatable, but involves taking synthetic hormones at a dose prescribed by a doctor. The dosage might need to be adjusted, depending on the demands of your schedule and other health issues.

Another distinctive feature of hormones is related to our "flight or fight" response. Our hormones regulate how we deal with stress. Therefore, when we feel stressed, our hormones kick into high gear to find a way to reduce the stress level.

Every cell in our body is exposed to hormones due to the circulation of our blood throughout the body. However, a hormone can only really affect a target cell that has the proper receptor for that specific hormone. Once the hormone bonds with its receptor, the proper biological response is triggered within the target cell. Hormones eventually are excreted from the body after they have been broken down.

As we mentioned previously, the process of hormones binding with their target cells can be disrupted. One way that occurs is because the hormone disrupter binds into the receptor instead, blocking the hormone from making its connection. This disruptor also triggers a false signal, which can cause that cell to follow the wrong set of instructions. There are multiple medical issues that can result from these disrupters.

For example, these disrupters can alter how much hormone is made, how fast the hormone degrades and as we mentioned, the way the target cell responds to the false instructions. Major issues can include disrupting the development of an embryo or altering the functions within an adult body. There have even been consequences such as cancer or other diseases as a result of hormone disruptors.

Therefore, it is important for scientists to continue the research to understand the long term consequences of hormone disruptors. When they send false signals, the body may have only have to take a small period of time to repair the damage. The brain also is attempting to return the body to homeostasis, since these disruptors can provide false signals that have a negative effect in that area. However, when longer exposure occurs, the body may be unable to completely repair itself or reduce the effects of the damage incurred.

Understand that this messenger system is constantly in motion. The brain uses hormones to find out what is going on at a cellular level with the body. At the same time, the brain communicates with the hormone producing glands to send out increased amounts of hormones during specific times, such as increased growth or stressful situations.

Growth Hormone

Another specific type of hormone is the growth hormone. This particular hormone is the focus many sports investigations.

Many articles have been written that argue increasing your growth hormone can improve muscular hypertrophy. At the same time, it has also been associated with decreasing body fat, while holding back the aging process. Some people in the body building industry have started taking these hormones in a supplement form to achieve these benefits.

But does this growth hormone really have these almost miraculous properties? To understand what growth hormone can do and what it cannot, we need to understand how it is made and the original purpose of this particular hormone.

Growth hormone is made in the anterior pituitary gland. This hormone is released in a pulse format. The pulses are controlled by another gland, known as the hypothalamus. The hypothalamus sends signals to shut down the release of hormone or to initiate a pulse that releases growth hormone into the body's bloodstream. When the body reaches its limit of growth hormone, there is a messenger known as the somatostatin that tells the anterior pituitary to shut down its releasing of hormone. The reverse is initiated with the growth hormone releasing factor (GHRF), which is another messenger made to initiate the necessary pulses.

Obviously, the role of growth hormone is initiate growth within the body. But how does this happen? The hormone has several different mechanisms available. The first is to tell the body to increase protein synthesis. The second is to increase lipolysis, which tells the body to decrease its fat load. The third is to increase sarcomere growth, a part of the connective tissue throughout the body. The fourth is to reduce liver intake of glucose but an enhanced formation of new glucose. Finally, growth hormone supports the pancreas and its own hormone production. The pancreas is known to produce insulin, ghrelin and glucagon.

What this all means is that growth hormone initiates and supports the growth of tissue in a variety of ways and decreases the adipose tissue. By maintaining optimal levels of this hormone, one can support not only a favorable body composition, but also overall physical well-being.

As we mentioned with other hormones, there are ways to disrupt the production of growth hormone. If you do not get adequate sleep, chronic stress, high blood sugar levels, a form of testosterone (DHT) and glucocorticoids. While you can increase your growth hormone production through low blood sugar and fasting, in the long run, these can cause extensive stress on your body and actually inhibit the production of growth hormone.

Yet, as with many hormones, growth hormone has specific effects on men and women. Estrogen also falls into this category. So how does growth hormone effect women versus men? Studies have found that women actually create more growth hormone than men. It seems that this increase can be related to the fact that women also have estrogen in their bloodstream. The combination seems to signal greater amounts of growth hormone in the body. Especially when women have exceptionally high levels of estrogen (think during the peak of their menstruation cycle), they also see peak production of growth hormone as well. Women also tend to spend a significant amount of their day actively producing and secreting the hormone, while men tend to have their largest pulses at night with minimal production and secretion during the day.

Women also tend to have a greater growth hormone response during exercise at all levels and they also tend to reach peak concentrations of growth hormone earlier in their exercise routine. Therefore, we can see how changes to our exercise patterns can increase our growth hormone production, especially for women. Speaking of women, let's move to a quick overview of estrogen, since it has a particularly strong tie to women.

Estrogen

While estrogen is most often associated with women, the truth is that women's bodies also produce testosterone and progesterone. The ovaries are the primary producer of these hormones for women. While they are fairly inactive during childhood and infancy, these glands kick into gear during fetal development and puberty. Once puberty is complete, women will settle into a cycle of ovulation, which also assists in regulating the production and secreting of these specific hormones.

Men also have the hormone estrogen, but again, these levels are much smaller than what would be found in a woman. Men have larger amounts of testosterone than women, although women also receive doses of it throughout their lives. As we will see, these hormones play specific roles for both women and men beyond what they are typically associated with. Let's learn what a few of the jobs of these hormones are and how they affect men and women differently.

Hormones are produced in cycles, which means that there are periods when these hormones are lower than others. Estrogen's primary jobs include building up the uterine lining for potential pregnancies, stimulates the building of breast tissue during puberty and also thickens the vagina wall. This hormone has also been associated with good bone and heart health throughout a woman's life. However, estrogen production in older women has been associated with increased risk of breast and ovarian cancer.

Progesterone works with estrogen to assist in the implanting of a fertilized egg in the uterine lining. In addition, this hormone also has effects on tissues that might also be sensitive to estrogen throughout the body. Finally, testosterone in women plays a role in increasing their sex drive, generating energy and the development of a woman's muscle mass. Men find that testosterone contributes to their sex drive and increasing muscle mass.

Menopause is a complete altering of the way that these hormones interact with the body. For most women, menopause means that the ovaries go into retirement. They are no longer producing eggs, so they also stop producing most of their hormones. While it does not happen all at once, the body does begin the process as early as a woman's thirties. When estrogen production ceases altogether, then women find themselves dealing with various health issues, such as hot flashes, lost muscle tone and even changes in a woman's sleep cycles. Therefore, women might find themselves having to make the decision to replace that lost hormone or to simply treat the symptoms that result from the loss with various prescriptions or natural methods.

A woman can choose to do a hormone replacement therapy to reduce these health concerns and minimize symptoms, but for those women who have a history of breast, ovarian or cervical cancer should discuss any potential hormone replacement therapy with their doctor. After all, it might not be as effective an option for those with high cancer risk. Most women find that menopause brings a whole new set of challenges in terms of their health, because they must weigh multiple pros and cons to all types of treatments.

As we have seen, hormones have a powerful role in the internal workings of the body. When they are functioning correctly, we can feel our best physically. Yet when they have been altered or disrupted in any way, there are consequences in terms of our health and overall well-being. Thus, we might find ourselves needing to visit a doctor or use supplements to allow our body to reach homeostasis.

Now let's discuss the other messenger system within the body, the nervous system!

Chapter Five:
The Nervous System

As we have discussed, the body has two main messenger systems that allow the brain to send instructions to the various organs, glands and muscles throughout the body. While hormones flow through the bloodstream to reach the individual cells to transmit the necessary instructions for a variety of tasks, there is another method available to the brain for transmit commands. That secondary option is the nervous system.

So how does the nervous system work? What are the parts of the nervous system and how do they work with the brain itself? Let's find out.

The most important part of the nervous system is the brain itself. This is the powerhouse of the whole nervous system and the control center of the whole body. The nervous system itself has both voluntary and involuntary actions. One of the primary parts is the central nervous system (CNS). The CNS is made of the combination of the brain plus spinal cord. A peripheral nervous system (PNS) is primarily made up of nerves that course throughout the body. These nerves are enclosed in bundles of long fibers, which are known as axons.

These axons essentially provide a connection from the CNS to our whole bodies and all our organs. While messages must be transmitted from the brain to the body, the body needs a way to respond. Therefore, the nerves transmitting the brain's messages are called motor or efferent nerves. The nerves that return messages to the brain are called sensory nerves. Most of the bundles of nerves actually perform both functions, thus the term mixed nerves.

Now the PNS is also divided into three parts, the somatic, autonomic and the enteric. Somatic are in charge of voluntary movement. The automatic deals with the involuntary aspects of the nervous system. It can also be separated again into the sympathetic and the parasympathetic.

Sympathetic nervous system is particularly in high gear during emergencies to mobilize the body to get energized. The parasympathetic nervous system is active during the relaxed state. The combination of these two systems allows for the individual's flight and fight responses, in addition to many other functions that happen throughout the day. Consider this part of the nervous system as the background application that is always running behind the scenes, but with the ability to notify the main system when necessary.

Finally, the enteric nervous system is in charge of the gastrointestinal system. This also functions involuntarily as it controls digestion and the movement of food through that system.

But to truly understand how this system functions throughout our bodies, it is important to learn about the nervous system from the cellular level up. Let's start with the types of cells often found throughout the nervous system.

Cells

The nervous system contains two types of cells, the glial and neurons. The neuron's fundamental property is how they communicate with other cells using synapses. These are membrane to membrane junctions that allow for rapid transmission of signals through electrical or chemical means on the molecular level. Neurons may also include an axon, which allows for thousands of potential synaptic contacts.

Before you start to think that neurons are cut from the same cloth, you need to understand that they exist in a large variety with different functions. Some of these neurons are sensory neurons that take physical stimuli into neural signals. Another type of neuron is the motor neuron, which transmits neural signals into action, either by the muscles or the glands themselves.

The glial cells provide support and nutrition, form myelin, work to maintain homeostasis, and participate in signal transmission within the nervous system. For example, within a human brain, it is estimated that the total number of neurons is roughly equal to the number of glial cells. When it comes to describing the job of the glial cells, the best way to explain it is that they are the support staff for the neurons. Not only do they work to provide nutrition and keep the neurons in place, they also assist in fighting pathogens that might attack the neurons, as well as cleaning out the dead neurons.

However, the most important function of the glial cell is the creation of a fatty substance called myelin. This substance wraps around axons to provide electrical insulation to allow for rapid transmission of action signals. When myelin is damaged, it is not necessarily able to regrow, so the damage can be permanent. There are diseases that actually strip myelin away from neurons, trapping an individual in their own head. Treatments continue to be researched to find ways to fix this problem for those with specific genetic disordered.

Now let's move up to the chain to have a better understanding of the central nervous system (CNS) and peripheral nervous system, known as the PNS.

Central Nervous System and Peripheral Nervous System

As we have mentioned, the CNS includes the spinal cord as well as the brain. The spinal canal is the home of the spinal cord, while the brain is housed with the cranial cavity. The protection for the CNS, the body has the meninges, which is a three-layered system of membranes. One of these layers is the outer layer known as the dura mater. The skull also protected by the skull, made of a hard and strong bone material. The spinal cord also has protection in the form of vertebrae, another type of strong bone. This is the part of the nervous system that most of us are familiar with.

The PNS describes all the other aspects of the nervous system that are not included in the CNS. Axon bundles are considered within the PNS, even though there are some axon bundles that can be found within various parts of the CNS, including the brain and spinal cord.

The PNS can be divided down even further, into two parts known as the somatic and visceral. The somatic includes nerves associated with the joints, skin and muscles. Dorsal root ganglia are considered the cell bodies of somatic sensory neurons. The visceral is often referred to as the autonomic nervous system. It contains neurons that deal with the internal organs, glands and blood vessels.

Now the visceral can be divided again, into two more parts. When mapping this out, it can sometimes appear similar to a long family tree. The two parts of the visceral are the sympathetic nervous system and the parasympathetic nervous system. The sympathetic nervous system is in charge of the body's fight or flight response. However, when it does not need to fire up our defense mechanisms, the sympathetic nervous system is active in maintaining the body's homeostasis. See the chapter on brain for a further discussion of homeostasis and how the brain maintains it throughout our bodies.

The parasympathetic system compliments the sympathetic by stimulating the feed and breed or rest and digest responses within the body. Thus, digestion and other internal processes happen automatically without our conscious thought to keep it going. Breathing is another activity where the body depends on the automatic systems to keep it going, especially as we sleep or our conscious becomes involved in other activities.

But while we have discussed a complex nervous system, most often found in creatures with vertebrae, there are other creatures that demonstrate variations of the nervous system. Scientists often refer to them as examples of simpler neural connections or alternatives to a complete nervous system. We will refer to them as neural precursors.

Neural Precursors

When you look at a sponge, you do not see the typical indications of neural pathways. In fact, the sponge has been classified as having no nervous system at all. The typical synaptic junctions are missing, thus there are no neurons within these creatures. However, messages must be sent throughout a sponge's body. So how is it done?

Sponges make groups of proteins that cluster together. When completed, the structure closely resembles a postsynaptic density, which is similar to the receiving end of a synapse. Yet the sponge's use of this particular protein cluster remains unclear. What is known is that sponges currently communicate using calcium waves and other types of impulses. This allows for simple actions, such as the complete contraction of the body.

However, let's face it. Sponges aren't really going many places. Their job is filter ocean water, thus feeding themselves and cleaning the ocean at the same time. But within the ocean are more complex creatures that also have examples of neural precursors.

One such example are jellyfish. These creatures have a diffused nerve net, instead of a central nervous system. The nerve net is typically spread throughout the body in an even fashion. Made of sensory neurons that are sensitive to visual, tactile and chemical signals; motor neurons that activate body wall contractions; and intermediate neurons that provide the communication between the other two. While this is a fairly unstructured nervous system, it clearly demonstrates that synapses occur in other creatures. However, the complexity of a human nervous system is unique due to the variety of messages that are constantly in motion through it. Messages on such a significant number of topics, from eating to comprehending information from the various senses and then acting on it.

So how does the nervous system manage all of this incoming and outgoing information? The answer is in a term called bilateria.

Bilateria

For a vast majority of animals with a nervous system, there is a division between their left and right sides. Simply put, the right and left side are almost mirror images of each other. Each side had its own cord or ganglion for each side, with the largest section of the ganglion being up front and often referred to as the brain. Our nervous system follows a similar pattern.

The spinal cord has segmented ganglia throughout the body that give each section of the body access to motor and sensory nerves. These segments feed into the main trunk and send messages to the brain. Depending on the creature, the main nerve cord will be either on the bottom or top of the body.

There are a variety of species, each with their own variations of this type of nervous system. The more complex the nervous system, typically the better developed sensory organs. For many arthropods, which is a group defined by insects and crustaceans, this means they have compound eyes and antenna. With these advanced sensory organs, they are able to process a variety of information from the world around them. We see this with flies all the time, as they quickly move at any sign of danger due to the abilities found in their compound eyes.

As we have seen, nervous systems can be more or less complex, depending on the creature and the complexity of its organ structure. For instance, a jellyfish's neural pathways are much less complex than the one found in a higher functioning mammal, including humans. But when we come down to it, the nervous system has some pretty basic functions, which we will explore next.

The Functions of the Nervous System

When we analyze it, this system is a major source of communication within the body. At its most basic level, it is a system that involves sending signals from one cell to another, creating chains that allow for messages to be transmitted from one area of the body to another.

What makes this stand out from the hormone type of messaging is its point to point signaling process. Think of it this way. When the government wants to get a message out to a large group of the population, they use a broadcast system. This spreads the message through multiple forums, but there is no guarantee that everyone will hear the message.

However, when someone receives a phone call, the message is clearly directed at them or targeted. The result is that there is confirmation the message was received. Our nervous system works much the same way. While hormones provide more of a broadcast type of delivery, the nervous system provides much more direct messages to specific targets throughout the body.

Not only is it targeted, but it moves much faster than a hormone message. Scientists have found that the fastest nerve signal will move through the body at speeds exceeding 100 meters in a single second. This is faster than just about any other type of movement. As a result, the body is able to quickly make changes and adjustments throughout our entire system in real time with minimal delay.

Yet it is so much more than a speedy message system. This is meant to control the body, both by acquiring information from the surrounding environment, but also then processing the information to find the appropriate response. Then messages are sent to various parts of the body to facilitate the response action. What is amazing about this process is that it happens so quickly. Most of us do not even think twice about how much work goes into the brain's translation of the input from our eyes, ears and touch.

As we have seen, signals are sent via the axons in either a chemical or electrical signal. While electrical synapses make direct and specific connections, chemical synapses are the more common type and are also more diverse in their abilities.

On the molecular level, these cells have specific receptors and transmitters that facilitate all of the signals that flash throughout our bodies every second. But while all this is fascinating, there is one aspect of the nervous system we have not truly explored. Yet it deserves a special discussion because of how critical it is. Yes, we are talking about the control center of the nervous system and our bodies as a whole, the brain!

Chapter Six:
The Control Center: Our Brain

Every human and many creatures have some version of the brain. It houses the judgment center, the decision maker and the overall control over an organism. Throughout this chapter, we will be learning about the parts of the brain and the various tasks the brain completes on a daily basis. While many of these functions and parts might be translated to a variety of creatures, for the purpose of this chapter, we are primarily going to be referencing human beings.

<u>The Parts of the Brain</u>

The size and shape of every brain is slightly different. This also means it is not always easy to find the common features from one brain to another. Yet, scientists have discovered some interesting aspects of brain architecture that seems to apply no matter what type of species. So we are going to start looking at various parts of the brain, starting with the cellular level.

As we discussed in the chapter on the nervous system, the brain is made up of the same type of cells that one finds in the nervous system. These cells are the glial cells and neurons. Glial cells perform a variety of functions, such as metabolic support, structural support, guidance of development and insulation. While this is all an important part of the brain, the most critical cells are considered the neurons. Why is this the case?

Because these cells are the ones that have the ability to target specific cells across long distances, allowing for messages to be relayed to all the organs of a body.

There are six main regions within the brain. Below we will discuss each of these areas and any associated functions.

- Telencephalon – Otherwise known as the cerebral hemispheres or cerebrum; it contains the cerebral cortex. This region is made up of two cerebral hemispheres and their various cortices, as well as the outer layers of gray matter and the internal layers of white matter.

- Diencephalon – Part of the forebrain; Consists of the hypothalamus; thalamus; epithalamus and subthalamus. Optic nerve attaches to this region. Within this region, there are multiple glands that control hormone delivery throughout the body, as well as control of visceral activities within other areas of the brain and the automatic nervous system.

- Mesencephalon – Known as the midbrain, associated with hearing, vision, sleep/wake, arousal, motor control and temperature regulation. This area is located below the cerebral cortex and above the hindbrain, so it is typically found in the center of the brain. Contained within this area is the tectum, inferior and superior colliculi, cerebral peduncle, midbrain tegmentum, substantia nigra and crus cerebri. Considered part of the brainstem, this region is associated with motor system pathways, as well as playing a role in the motivation, excitation and habituation, while relaying information for hearing and vision.

- Cerebellum – Tightly folded layer of cortex, include four deep nuclei. Three lobes can be defined; the anterior lobe; the posterior lobe; and the flocculonodular lobe. In charge of precise motor control; calibrating the detailed movement's form but it does not decide which movements to execute. Involves feedforward processing, because there is little internal transmission between input and output during the signal processing. Also plays a part in motor learning, particularly if there are fine adjustments that must be made during the motor function.

- Pons – Part of the brainstem. Lies between the midbrain and the medulla oblongata, but is in front of the cerebellum. Regulates the change from inhalation to exhalation. Plays a role in generating our dreams. Deals with respiration, sleep, swallowing, hearing, bladder control, equilibrium, taste, facial expressions, posture, eye movement and facial sensation. Divided into two parts. Often associated with touch and pain for the face.

- Medulla oblongata – Located in the hindbrain. Cone-shaped and responsible for various involuntary functions, including vomiting and sneezing. Contains respiratory, cardiac, vasomotor and vomiting centers. Deals with heart rate, breathing and blood pressure. Connects higher levels of the brain to the spinal cord. Functions include regulating various reflexes, including the swallowing reflex.

Throughout these various regions, there are multiple smaller areas and parts, including various nerves, that make up each of these distinct areas. There is plenty of overlap between regions, but the most important thing to take away is that the brain works harmoniously, despite all the various regions that must be part of the process to get anything accomplished. The other important takeaway is that most of these things happen nearly instantaneously. It takes much longer to describe these functions than it does for the brain to actually execute them.

As we have discussed, there are multiple ways for these messages within the brain to be relayed. Ultimately, these processes occur on a cellular level and are the building blocks of the other major components of the brain. Keep in mind, the components are smaller parts of these regions and these components have smaller parts that are part of them. The brain is very complex, as we will continue to see. Below is information about each of the components found in the brain:

- Cerebral Cortex – Outer layer of gray matter found in the cerebrum; generally classified into four lobes: the occipital, frontal, parietal and temporal.

- Medulla – Contains small nuclei, which along with the spinal cord, contain a wide variety of involuntary motor functions, including vomiting, digestion practices and vomiting. It is also a source of sensory functions as well.

- Pons – Located within the brainstem, it is directly above the medulla. It controls voluntary but simple acts, including respiration, bladder function, eye movement, posture, facial expressions, swallowing, sleep and equilibrium.

- Hypothalamus – A region of the brain at the base of the forebrain, it is engaged in involuntary or at least partially voluntary actions. Some of these are eating and drinking, release of some hormones and sleep/wake cycles.

- Thalamus – Responsible for relaying information and motivation back and forth between cerebral hemispheres. This also includes the subthalamic area, which is defined by its action generating systems, including behaviors such as eating, defecation, drinking and copulation.

- Cerebellum – This is the modulation center for the outputs of brain systems, regardless of whether they are motor or thought related. Muscle coordination is an example of how this part functions. While it does not provide instant precision, this is where we really do our learning. In fact, this part accounts for almost 10% of the brain's total volume and at least 50% of all neurons are held within this structure.

- Optic Tectum – This part controls actions that are directed toward various points within space, most commonly related to visual stimulus. It also directs movements involving reaching and other object-directed acts. While it does receive visual input, this area also receives inputs from the other senses when it relates to directing various actions. It is also identified as part of the midbrain area.

- Pallium – Found on the surface of the forebrain, this layer of gray matter is involved in smell and spatial memories. However, this area of the brain consists of multiple folds that increase the surface area of the pallium. As a result, the amount of information that can be processed and stored increases with the increase of folds.

- Hippocampus – Only found in mammals and is involved in complex events, such as navigation for various animals.

- Basal ganglia – An interconnected group with the forebrain. Primary function is to make action selections and this area can send inhibitory signals throughout the brain to stop various motor behaviors. Reward and punishment have the greatest effect in this area, by altering connections in this part of the brain.

- Olfactory bulb – Processes olfactory sense signals and then sends the output to the pallium for combining with other sensory information. Smaller in humans than other vertebrates, although primates also have a smaller bulb.

As we have seen, the brain is made up of multiple parts. While it seems that there is some cross over in terms of what each area is responsible for, this redundancy also makes it possible for the brain to bounce back from various injuries because other areas can be reassigned to take over functions for the area that has been damaged.

When looking at brains across a variety of species, one thing that becomes clear is that size and shape do play a part in the processing power of one brain versus another. For instance, the size and shape of forebrain result in dramatic difference in terms of what a creature might be able to do with their brain. Now let us spend some time learning about the development of this amazing control and processing center.

Brain Development

The development of our brains is something that fills many individuals with awe. This complicated and high functioning organ starts as a simple swelling of the nerve cord before growing into the complex organ that gives us the ability to do so much, including appreciating art, understanding science and mathematics, as well as have feelings, morals and ethics.

As the brain develops during the embryonic stage, it builds an extremely complex array of connections through different areas. Neurons are created by stem cells in specific zones, then begin their migration to reach their final destinations. Once they are in position, they begin to sprout axons, navigating through the brain via branching and extensions. Eventually, they reach their targets with the tips of their branches and these tips form the synaptic connections necessary for the brain to exert its control over the body.

Excessive numbers of these neurons and synapses are created, but the unneeded ones are eventually pruned away.

Yet this growth and creation of the brain and nervous system requires more than just the creation of a few synapses. During the embryotic stage, a narrow strip of ectoderm is left running down the back. Eventually, this narrow stripe will become part of the new spinal cord housed in the spinal column in your vertebrae.

This stripe of ectoderm first becomes the neural plate, which eventually folds inward to create a neural groove. The lips of this groove merge together to enclose the neural tube. What is a neural tube? It has been described as a hollow cord filled with cells and a fluid filled ventricle, which can be found in the center of the neural groove. The front of this tube swells to form the parts that will eventually become the forebrain, midbrain and hindbrain.

Now that we have the early stage of the brain, big changes start occurring. The bulge that forms the forebrain begins to split into two vesicles. These will end up being the telencephalon and the diencephalon. The hindbrain also splits in two parts, the metencephalon and the myelencephalon. All during the process, those neurons are branching out and making more and more connections. Still, those connections are on a cellular level, contributing the building blocks that become the gray and white matter that will end up folding into various parts and regions of the brain. Other cells will step in and blood vessels to feed these cells will also be created. In nine months, this amazing organ will be ready to start learning, while continuing its growth pattern. While glial cells will continue to be produced throughout our lifetimes, neurons are primarily created during pregnancy and early childhood. As adults, we are limited in our ability to have new neurons generated. As a result, there can be irreversible damage to the brain when a large number of neurons are killed off.

There have been many debates about nature versus nurture when it comes to brain health and development. However, studies seem to suggest that both play a role. While genes may determine the shape of the brain and how it might react to experiences, the experiences themselves refine the matrix of synaptic connections within the brain itself.

In the next section, we will examine how the brain's electrical and chemical processes work during the process of making those connections.

<u>It's Electric – Understanding Electric and Chemical Activity in the Brain</u>

Within the brain, there are constant neurotransmitters at work. These are chemicals that get released at various synapses when the possibility of action actives them. Those neurotransmitters attach to special receptor molecules, modifying the electrical or chemical properties of those receptors.

Neurons release the same transmitter or combination of transmitters to all its synaptic connections. So neurons can be classified by their transmitters. Psychoactive drugs work on the principle that they are altering neurotransmitter systems.

There are two types transmitters that are found most frequently in the brain. One is glutamate and gamma-aminobutyric (GABA). Glutamate is known for its excitatory properties on target neurons, while the gamma-aminobutyric is associated with inhibitory effects. For example, anesthetics reduce glutamate's effects, while GABA is enhanced by means of tranquilizers. As a result, it is important to remember that each neurotransmitter has its own distinct properties and those properties can be altered through a variety of substances.

Other chemical neurotransmitters are available within the brain, but they are used in some very limited areas and are dedicated to very specific functions. An example is serotonin, which comes exclusively from a very small are of the brainstem known as the Raphe nuclei. Serotonin is the target of diet drugs and antidepressants. Another transmitter known as norepinephrine is responsible for arousal and produced in the locus coeruleus, which is also around the brainstem. There are two other transmitters that have multiple sources within the brain, acetylcholine and dopamine, but they are not as widely distributed as GABA and glutamate.

As we can see, the brain sends messages via a chemical function, but if the right combination does not occur, the messages will not get sent. Taking in drugs and other substances can also alter the type of message that is sent. For instance, we see that individuals taking in large amounts of alcohol or other drugs can change the production of different transmitter, resulting in a lowering of inhibitions and blocking the judgement center of the brain from doing its job.

Yet, as we have learned from other sections of the body, humans have a certain amount of electrical charge. In fact, we create electrical fields. This is also true within the brain. There are multiple electrochemical processes and the side effect of them is the electric fields they generate. If a large enough number of neurons are firing at the same time, the electric field they create can actually be detected outside of the skull. Medical professionals use an EEG or MEG machine to record these electric fields. The results have shown scientists that even while we are asleep, the brain is constantly active.

Imagine a machine that is constantly humming, processing billions of pieces of information in a matter of seconds. This is what the brain is constantly doing throughout the day. At night, when we are resting, the brain is directing traffic within the body. Tissue repairs, muscle repairs and other necessary functions that allow our bodies to function throughout the day are completed while we sleep. This is the time when the brain is supposedly at rest. When we are awake, there is even more for the brain to complete, process and make decisions about.

So what does all of this activity look like on an EEG? Simply put, it appears as a variety of wave lengths. Delta waves are large and slow, appearing most frequently during sleep. Alpha waves are a faster version of delta waves and they might appear slightly shorter. However, it has been noted that alpha waves appear when we are awake, but not necessary attentive on a specific task. The charts look chaotic with spikes when we are actively engaged in a task or project that has our full attention.

Yet when a brain is damaged by seizures, the EEG can show a chart with almost pathological levels of electrical activity. This is because the brain's inhibitory mechanisms are not functioning. As a result, the pattern shows large waves and various spike patterns of an unhealthy brain. A seizure, in effect, causing the electrical timing of the brain to be off and so the brain struggles to realign itself after a seizure is over. This is a particular area of study for researchers, because it is important to understand both how the electrical activity effects the brain and how to create medicine that assists in the process.

Throughout this basic study of the brain, we have learned about just a few of the complex processes that occur every second. It not only assists us to make decision and be insightful, the brain is also directing all the other myriads of functions occurring within the body itself. This includes fine motor movements, breathing and digestion, just to name a few of the processes.

Within our brains, however, there is another unique process that occurs apart from any other in our body. It is metabolism.

The Brain's Metabolism

The body has a metabolism, which is the rate at which molecules are either prepared for storage or used for a specific immediate use. Molecules can also be made into by-products through these chemical processes. For example, our bodies use metabolism to either create or burn fat. While it might be easy to assume that metabolism is constant throughout our bodies that is not really the case.

Our brains actually run at a different metabolism that is apart from the rest of the body's metabolism. This makes sense if you think about it. The brain must do so much processing and movement of messages throughout just a few minutes that it cannot afford to have a metabolism geared to the same rate as the rest of the body.

Glial cells control the chemical composition of the brain's fluid, even down the ion and nutrient levels. The reason for this particularly high level of control is that the brain is constantly consuming large amounts of energy, even though the brain itself is relatively small. Most of this energy is used to maintain those electrical charges that we discussed in the last section.

Therefore, the brain maintains a separate metabolism to adjust for its very unique needs. While most of its energy comes from glucose, the brains also use ketones, fatty acids, lactate and acetate. There is still some discussion about whether or not amino acids can be added to this list as well. One final area we will discuss is the information processing capabilities of the brain.

The Brain: Our First Personal Computer

When we think of information processing, our first thoughts go to our computers. Most of us own a version of a smartphone, which is literally a hand-held computer. The processer in our phone can handle multiple tasks at our request, including making a call while getting us directions to our destination and notifying us that our best friend just posted to Instagram.

Yet the processing power of this one small device is not even a portion of the processing power found in our own heads. So how did this whole understanding of the brain as a processing center get started? In the 1940s, computers and mathematical information theory were being developed. Scientists looked at the way these machines were operating and they wondered if the same patterns could be applied to the brain, thus giving us a better understanding of how it functioned.

As a result, a new field of study known as computational neuroscience was created. Simply put, the processing of the brain was originally thought of in terms of algorithms and the flow of information. Science, however, has begun to use collected data to move into a more realistic understanding of how it all functions.

One of the early contributions came from a paper that examined how the various visual responses of the retina's neurons and optic tectum of frogs. The study concluded that the neurons in the tectum essentially combined elementary responses. This study was completed with frogs and they found that the tectum became a bug perceiver that allowed the frog to find food.

Other investigations have led to a growing body of knowledge that shows how the brain is able to produce increasing complex responses, despite distance between the various parts of the brain. Cells can have a wide variety of responses to what we see, even if they appear to be unrelated to the actual sense of vision, including memory responses and abstract cognition types.

To understand how all these response patterns work, scientists have begun to put together mathematical models, using a variety of powerful computers. Some of these models are abstract, meaning that they focus more on the concept of neural algorithms versus how they are actually implemented in the brain. Others try to use the data to better understand the biological properties of the actual neurons themselves.

Still, none of these models have been able to completely capture the full function of how the brain operates.

Then you have to understand that these models are based on what groups of neurons are accomplishing. When you get down to a single neuron, you find something that is unique and incredibly complex. These neurons are able of completing the equivalent of multiple computations on their own. Current methods of computing brain activity can isolate a few dozen neurons and their actions at best. Clearly, there is much more that will be learned about the brain's processing power as time goes on.

The Human Brain Project has begun to the process of building a realistic computational model of not just one area of the brain, but the whole thing. Whether they succeed or not remains to be seen, but it is clear that they are moving in a direction to better understand this unique control center and how it actually manages to do all that these neurons and cells do every single day. Throughout the next few paragraphs, we will discuss a variety of functions completed by the brain that involve multiple areas of the brain. As you will see, these working relationships are precise and extremely accurate, guaranteeing that what needs to happen does so, every time.

One of the first areas of brain function that is of interest to scientists is how the brain processes all the sensory information it receives. Remember, the brain is not just receiving visual or auditory signals. It is receiving both and so much more from the other senses. All this information needs to be combined into one complete picture. Thus, the brain takes in all these signals and processes them, before then sending the signals to the area of the brain that handles that particular signal. Eventually, everything is reintegrated and the brain makes decisions and prompts actions based on that information. What scientists are eager to understand is how all of that information is processed at such high speeds on a continual basis.

There are three further areas of the brain that we will discuss before we close this chapter. The first concept is movement. There are multiple areas involved in the process of movement. Below are the major areas that scientists have identified as critical to movement.

1. Ventral horn – This part is located in the spinal cord. It contains those neurons related to activing our muscles.

2. Oculomotor nuclei – This part is located in the midbrain. It contains neurons that active the muscles related to our eyes.

3. Cerebellum – This is located within the hindbrain. It is the calibration system of the brain that assists with the timing of movements. Often associated with fine motor control.

4. Basal ganglia – Located in the forebrain, this part of the brain chooses an action based on motivation.

5. Motor cortex – Located in the frontal lobe, this part directs cortical activation of several of spinal motor circuits.

6. Premotor cortex – Also located in the frontal lobe, this area is in charge of grouping movements into coordinated and sometimes complicated patterns.

7. Supplementary motor area – Located in the frontal lobe, this area controls and sequences various movements into temporal patterns.

8. Prefrontal cortex – Located in the frontal lobe, this area is essentially the boss of the operation. It is charged with various executive functions, including planning.

In addition to all of these areas, there is extensive circuitry meant to control the ANS, which is involved in secreting hormones and also modulates all of the gut's muscles. In addition, the ANS affects or has input on our heart rate, salivation, perspiration, digestion, respiration and even sexual arousal. Most of these are not considered directly under our control. Yet the brain is aware of them and has accounted for them.

Pretty incredible, isn't it?

As we mentioned, we were only going to cover three more topics. One was movement, but the second one is arousal. When it comes to behavior, the one most easily pointed to as common across species is sleeping and waking. While most of wish morning did not always come quite so early, the reality is that every creature follows some type of sleeping ritual. Bats, for instance, sleep during the day and do their insect hunting at night. This is their particular sleeping ritual. Humans are programmed to sleep when it gets dark and be active during times of light.

To be able to keep these rituals effectively, the brain must have some type of internal clock. Not only does it have this internal clock, but it is one tuned to a superfine scale, due in large part to the number of brain areas involved in this particular network. One of the key parts is the suprachiasmatic nucleus (SCN). This is a tiny part of the hypothalamus and it is the body's central and main biological clock. Activity levels in this clock rise and fall according to 24 hour periods, creating what is known as circadian rhythms. The rhythmic changes that occur in this area appear to be tied to what scientists refer to as the clock genes.

While you could take it out of the brain and it would still keep time, the SCN also takes in information from the eyes, so that it can calibrate itself against the light and dark we see every day. Therefore, the sunshine that we enjoy is also serving the dual purpose of helping our internal clocks to reset or realign themselves.

There are other areas that the SCN then projects to, thus resulting in the sleep/wake cycles we are so accustomed to. One important receiver is the reticular formation. Located in the lower brain, it signals the thalamus to send activity signals throughout our cortex. When this area is damaged, it can often result in a permanent coma for the brain because it simply does not know to wake up.

As we might know, sleep is a restorative state for the body. The brain directs a variety of activities while we sleep to repair and assist the body's growth. REM sleep includes dreaming, while NREM does not usually have any dreams involved. Our brains actually switch between these patterns throughout the night. Without both, we do not get the right amount of rest necessary to function. Chemicals that are active during our daily activities drop dramatically during sleep, including serotonin and norepinephrine. Acetylcholine is the opposite and tends to rise in quantity during sleep.

Although it seems a time when brain activity would diminish, we see that it in fact is maintained. In some areas, it actually increases. Another interesting fact is that we have our own internal clock that is more accurate than many other clocks, yet it never has to be wound up or reset. Our bodies calibrate it every day when we open our eyes. Truly, it is amazing how this one small organ controls so much.

Finally, we are going to discuss homeostasis. First of all, we have to understand what that really is. The process of homeostasis is based on the idea that there are specific parameters that the body needs to have maintained in order to survive. While some variation can be tolerated, the range is typically limited.

So what are these parameters? The list includes temperature, salt concentration, water content, blood glucose and even our blood oxygen level. While there are other parameters, these are considered the biggies. The brain, of course, makes this one of the highest priorities or a critical function. The basic idea is that any time one of these parameters shifts out of its optimal phase, a signal is sent to create a response that will return the parameter to that optimal state. This process is known as a negative feedback loop.

To see an example of this process outside of our bodies, you can look at your home's thermostat. As long as the temperature is maintained at the range you have set, then the furnace will not turn on. But if the temperature dips, the thermostat notes this variation and sends a signal to the furnace. The furnace then kicks on to pump heat into the home, hence providing the response that returns the parameter or temperature back to what you have determined is the optimal setting.

Our bodies function the same way. The part of the brain that plays the greatest role in maintaining our personal homeostasis is the hypothalamus. While small, it is clearly in charge of many critical functions throughout the brain and body. There are several nuclei within the hypothalamus that receive signals from sensors located in our blood vessels. These sensors report on all the parameters we previously mentioned, as well as giving the hypothalamus a location of the variation.

As a result, when the negative feedback kicks in, the hypothalamus turns on the output signals and attempts to address the issue or deficiency. Outputs may also go to other glands in the body, including the pituitary gland so that hormone messengers can also be sent out to create changes at the cellular level.

When we are ill, these parameters might be dramatically affected. Thus, the need to make changes at the cellular level, including sending out an increased amount of defensive white blood cells. Yet the goal of all these actions is to return our bodies to their happy place. When it cannot be achieved, the brain will send out additional signals and begin other actions meant to result in a return to homeostasis.

While this has been an extensive chapter, it was also a relatively short tour of all the functions that our brains accomplish on a daily basis. It may seem simple to close our eyes and fall asleep, but there is so much more going on behind the scenes. Our brains are changing wave patterns, sending signals to make repairs and of course, continuing its assessment of our body's well-being.

Throughout this chapter, we have learned about many of the various regions and parts of the brain. Each part is not responsible for just one part or one action, but often have to demonstrate amazing multi-tasking capabilities to process all the signals that are coming in and then send out the proper output.

We have also covered a variety of individual glands that provide hormone messengers. As a result of these messengers, our bodies receive many necessary instructions from our brains. They also continue to maintain the overall parameters necessary to sustain our lives. This includes electrical fields created by the brain's activities and the areas of the brain that control movement.

Throughout this process, we have also learned that although the brain is very complex, this also provides a protection for us as individuals. The brain is able to repair itself by transferring jobs to other areas. Essentially, a new part of the brain can relearn the task that was previously handled by the damaged area. We see this in many individuals who have strokes that cause damage to one side of the brain or the other. Yet through physical therapy and effort, our brains can make the proper adjustments.

This is just a small overview of our brains, but it should leave us in awe of how much this small organ is able to do every day. While we might sleep, our brain never shuts off. Instead it continues to work and maintain our lives. What an amazing little organ!

Chapter Seven:
Biological Life Cycles

The lifecycle of an organism is defined by some very specific steps. Each organism, no matter who or what they are, follow these basic patterns. It must be noted that on the cellular level, new organisms are created through a cycle of cell division. But once we leave the cellular level behind, it is quickly obvious that the lifecycle is a pattern that has been repeated over and over again from the beginning of time.

Step One – The Beginning

When it comes to the beginning of an organism, most of us think first of babyhood. Yet even before babyhood, there is a moment where the process of creating an organism begins. That is known as conception. Every creature goes through this process, although some ways are different from others. For example, humans have sex, thus giving the egg an opportunity to join with the egg and settle into the uterus to begin the process of growing into a baby. Female frogs lay eggs and then the male frog fertilizes them. These are just two examples of the process of conception.

After conception, the fertilized egg is kept in an environment where it can grow to a specific level of maturity. This maturity level can differ based on the organism. The resulting new lifeform may be born and ready to care for itself upon hatching or birth, as in the case of frogs and turtles, both of whom have limited care from their parents once they are born.

Instead, these creatures are born in such large masses because the survival rate for the infants is not high. But larger numbers of young increase the chances that more will survive. As with many different species, offspring survival often becomes a numbers and percentage game.

Others are born requiring more parental involvement in their early development. Birds are just one example. The babies hatch and then the parents are constantly working to feed and care for them. They are also training the babies on how to function as adult birds, such as feeding themselves, building a home and even finding their own mate. Eventually, these babies take their first flights and begin the process of starting their own families.

Humans must also be active in the early lives of their offspring, providing food, shelter and training. Yet as they grow and develop, the needs of these organisms' change. Parents find themselves altering their interactions with their offspring. Birds stop providing food for their young at a certain point to force their continued development, but not all of their young will survive to adulthood especially if they have not taken the lessons of their parents to heart.

Step Two – Moving from Childhood to Adulthood

As babies develop and progress, they move into a phase of maturing toward adulthood. In humans, this can be described as childhood and the teenage years. During this time period, a child's body develops and grows into its adult version. Hormones kick in and the body matures. Hair grows in unique places, which is an outward sign of the inward changes that will allow this body to eventually produce a baby of their own. This process takes years. Parents use this time period to train and guide their children in social norms, as well as teaching them how to provide for themselves and skills they will need to survive.

Other animals have a shorter maturation process. Deer often only spend one year in this maturing process. By their first spring, they are considered mature enough to fight for their own female and to father a child. Does also quickly mature. Yet no matter the animal, eventually, they all mature to the point of starting the reproduction process.

Another part of this particular time of an organism is the amount of energy needed. Food consumption grows to accommodate the needs of the body during its growth process. As anyone with a teenager understands, food becomes an expensive commodity as the amount eaten continues to increase during a process of extensive growth that requires production of a large amount of energy and a big supply of the building blocks for proteins and cells. Inputs of fuel to meet the demands of the body are typically never this high at such a constant level again within the lifecycle, with the exception of pregnant females. Again, the demands of reproduction and the growth of a new organism prompt increased demands of fuel.

Step Three – Reproduction

At this point, mature organisms look for a partner to reproduce. The underlying demand of a vast majority of organisms is to reproduce. It is instinctive. Yet humans are unique in the fact that they can have some control both on when they reproduce and if they reproduce. Fundamentally, the desire and need of most organisms is to create offspring.

Some use this to create a family unit, while others just create the offspring without an active role in creating a traditional family unit. An example is fish. These organisms lay eggs, fertilize them and then wait for them to hatch. While the parents may provide some initial care, for the most part, the baby fish are on their own.

When it comes to humans, the desire to build a family is more than just an instinctive reaction. It also has social and cultural significance. Basic processes of biology, including the production of food and shelter, have often been elevated into human culture. Throughout this period of reproduction and adulthood, the organisms pass on knowledge and wisdom, as well as acquire these things for themselves. Still, the organisms eventually reach the final stage of their lifecycle, but one that also includes renewal for the biological diversity of our planet as a whole.

Step Four – Maturing and Death

All organisms renew on a cellular level, but at some point, they stop the process. Thus, the organism itself begins the process of aging. Organs breakdown, bodily functions begin to slow or stop and then the organism's ability to sustain its own life gradually disappears. Within the animal kingdom, few animals make it to die of old age, as the natural order of predator and prey plays a factor in keeping populations in check.

Yet if an animal manages to avoid being prey, they will eventually die as their body ceases to function. Humans have built rituals around the ending of life within the body. There are burial services and other traditions of particular cultural significance, depending on the area and group of humans. Still, in the end, the body itself stops all metabolic processes and begins to decay. Within the plant and animal kingdoms, this decay allows the earth to reclaim the nutrients and thus provide these building blocks to the next generation of organisms.

To aid this process of breaking down dead organisms, there are specific bacteria, insects and animals that reduce the dead cells into the basic building blocks. What is known as rot or decay is a part of the lifecycle, thus recycling cellular material to produce new plants, animals and humans. Most of the decayed material is first reused by plants, which are ingested by other organisms as food sources. Thus, a bear that dies in the woods becomes nutrients for the trees, which provide food for squirrels and birds. Even when plants die or are destroyed, they also return important elements back to the earth.

Farmers understand the cycle of renewing the soil better than just about any other group. Over the years, they have learned how important it is to rotate crops, so different nutrients are used and also replenished as the plants are turned back over and into the soil to decay during the winter months. Additionally, they recognize the importance of allowing land to rest, giving the soil an opportunity to renew and regenerate itself.

Life has a cycle that constantly depends on a process of creation and destruction. As a result, the lifecycle itself is one that is constantly in all these states at exactly the same time, because every organism is part of the larger process of life and survival on planet earth.

While most of what we have discussed can be attributed more to animals and humans, plants also follow that life cycle. Yet they also have some very distinctive processes to create food for themselves. So next, we will discuss the process of photosynthesis.

Chapter Eight:
Photosynthesis: Feeding Plant Life

When it comes to plants, the process of creating new plants and feeding themselves is very dependent on other organisms. The very creation of a plant itself is dependent on sunshine, water and the nutrients found in the soil. But first, a seed needs to be fertilized. Here is a perfect example of the need to use other organisms to complete this process.

Step One – Fertilization

A flower will produce either the egg or sperm depending on the sex of the plant. Some plants will produce both, depending on their species. Yet in order to achieve conception, the plant needs to rely on other species to provide the connection between the egg and sperm.

Bugs and insects are often called upon to provide that connection. Bees are known for their critical role in plant fertilization. They often pick up the pollen necessary to fertilize the eggs. As they move from flower to flower, they pick up and deposit pollen dozens of times. To repay that kindness, the flowers produce a nectar that the bees use to feed themselves. Thus, plants play a role in the larger process of biodiversity.

These fertilized eggs are found in the form of seeds, which fall to the ground. Once in the soil, the seeds begin to grow, placing roots and then sprouting leaves and stems. At this point, all the tools for photosynthesis are in place.

Step Two – A Little Water and Sunshine

The plants collect sunshine by means of their leaves. They use this collected sunshine as the energy to produce sugars, lipids and other proteins. But how does this occur? The first stage involves taking the sunshine and turning it into chemical energy, which is stored in the form of sugar. But how does this occur? The plant takes in sunshine, carbon dioxide and water. These are combined to produce oxygen, water and glucose aka sugar.

The leaves are both the collection point for the materials needed to produce the sugar, oxygen and water, but they are also the point where the results of photosynthesis are released. Stomatas release the oxygen, while gathering the required carbon dioxide.

Here the roots step in and gather the only thing the leaves cannot, which is water. It is piped up through the stem to the leaves, which are the factory where the magic happens. Once the process is complete, the materials are released and then the process begins again.

So what are the parts of the leaves? The chloroplasts are the factory where the process of photosynthesis. Within the chloroplasts, there are several important structures. Chlorophyll is the green pigment that absorbs the necessary sunshine. The stroma is a dense fluid within the chloroplast that is the actual site of the conversion of carbon dioxide to the essential sugar. Grana is a stack of thylakoid sacs, which are flattened and are the site of the conversion of solar energy to chemical energy.

The end result of this process is that organic compounds are created, which are used by both the plant and other life forms.

Light reactions produce the energy and the dark reactions, which occur in the stroma, produce the sugar. As you can see, the process is necessary to help the plant survive, but to also produce the oxygen necessary for other organisms, such as animals, insects, fish and humans.

It is this essential process that leads us into the larger discussion of the biosphere. As we shall see, from the cellular to the biggest picture of the lives on earth, all these processes show how they are interconnected. One organism cannot survive without the functions provided by other organisms. Therefore, the beginning of a biosphere is the beginning of the discussion of interconnectivity.

Chapter Nine:
Getting to Know the Biosphere

The easiest way to define the biosphere is to lump all life on earth into one group. The area that contains all this life is divided into five different levels. The first level is the uppermost atmosphere. It is the most extreme end of the all-encompassing biosphere. The other end of the scale is the lowest most ocean depths. Between these two are three other levels, the atmosphere, hydrosphere and the lithosphere.

The lithosphere layer is made up of rock, soil and sand. Thus, it creates the outer solid layer of the crust of the earth. Within this sphere are the tectonic plates that are constantly in motion. As a result, earthquakes and erosion are just part of what happens when these plates are moving.

The atmosphere is a defined area of gas made of nitrogen and oxygen. The gas gets thinner as the further up into the atmosphere one goes. The atmosphere also contains the oxygen and carbon dioxide necessary for all life to survive. This is the main layer that supports the largest amount of life, although it can be found in all the layers of the biosphere. This includes plants, animals and humans.

Finally, the hydrosphere encompasses all the water on the planet. This is a very critical layer, as it includes an element that is necessary to virtually all functions within biology. No matter what the process, water plays a part. In addition, this layer includes water based lifeforms, such as fish, dolphins and whales. There is also plant life within this level.

These spheres connect with each other on a variety of levels. Processes on one level have an effect on other levels. Volcanic activity, for example, helps to push water into atmosphere where it ends up moving into the hydrosphere as precipitation. This is just one example of how the spheres interact. By virtue of these interactions, over 30 different chemical elements move through the spheres, cycling throughout the environment.

Yet going down out of these levels, one begins to see how the interactions between organisms also affects the spheres. Carbon is one example of these interactions. Most of the organisms on earth participate in cellular respiration, requiring oxygen to function. The result of the different chemical reactions is the production of carbon dioxide. Plants use photosynthesis to take that carbon dioxide for their chemical reactions and then turn it into oxygen.

Humans have not been kind to the plant world, as current deforestation numbers would suggest. Changes in land uses can result in differences in temperature and water distribution, which make other consequences including soil erosion. As a result, the balance of the biosphere in the area of carbon has been tampered with, among other systems and functions of the biosphere.

Human development plays a large part in how the biosphere operates. Nothing done in the environment is without consequences in another area of the biosphere. The various processes within the biosphere demonstrate how much each organism and methods are connected to each other.

When humans interact with the biosphere, there are real life changes that occur. But the negative or positive effects can be hard to measure. Scientists have determined that the biosphere is in a continual motion. Therefore, what is normal at one point may not be within a few centuries. The environment is not a static thing, but constantly on the go.

The consequences of changes to the biosphere are also felt in another critical way, which is the mass extinction of different species. The end of diverse species also changes the face of the biosphere and its different layers. So let us move to a short discussion of mass extinction and its long term effects on the diversity of the biosphere itself.

Chapter Ten:
What Caused Mass Extinctions

As we have discussed various biological processes, it becomes clear that there is an intense amount of interconnectivity between the aspects of the biosphere. At the cellular level, there is a symbiotic effect and it continues up throughout the chain of organisms all the way to the biosphere levels.

This interconnectivity can have positive effects, but it can also lead to massive upheavals within the species. The result of these upheavals can be minor or they can be very dramatic. The worst case scenarios often result in the end of a species. When dramatic changes occur to an environment very fast, the organisms may not always be able to adjust quickly enough to the changes.

What is often the result? The mass extinction of one or more species may be the final consequence. Tampering with the environment can have long range consequences, both for the organisms, as well as the plants. Often, these extinctions happen fairly quickly, but the recovery from them can take millions or billions of years. Additionally, the recovery does not mean that the lost species are ever returned. These extinctions provide a purging of the biosphere.

Scientists have concluded that there have been at least five of these dramatic changes that resulted in mass extinctions. Most have been a result of changes in the environment, primarily cooling. One of these is the purging of dinosaurs. Yet a majority of these were the result of natural corrections within the biosphere itself. So when did humans start having a greater effect on the biosphere?

To put it simply, the human population began to have the greatest effect when they became more organized as a people. For example, hunting alone results in the extinction of 15,000 to 30,000 species every year. Unsustainable use of resources alters the biodiversity of the planet and can reduce or eliminate a species habitats. When habitats disappear, the extinction of a species can happen very quickly. No one would disagree that when someone's home is destroyed, it can reduce their chances of survival.

Let us think of it in terms of a domino effect. When one domino is pushed over, the chain reaction begins. In the same way, pushing one domino, such as deforestation, has also begun a chain reaction with much larger consequences.

Most species struggle because their homes are being destroyed far too quickly, giving the species no time to adapt to a new habitat. The overarching environment is attempting to adjust to all the changes being made by humans and as a result of human activity. Yet the adjustments are very dramatic, causing various conflicts between nations fighting for the most basic of resources, such as water and food. The resources that they do have available are often quickly depleted, without thought of sustainability or long term effects.

The nations that are not as developed find themselves trying to support growing populations without the access to the necessary assets. They are also ill-prepared to deal with the changes in the climate and the resulting significant weather events. Many would argue that humans did not create all of this change, that some of it is part of the natural progression of the biosphere. Yet even if that is true, humanity does create a footprint on the biosphere, for good or for evil. Our actions do have consequences. Some of these consequences are being felt now but will also be felt by future generations if corrections are not made to the course humanity is currently on. Symbiotic relationships are in jeopardy, not just for individual species, but for life on the planet as a whole.

Throughout the last several decades, humans have begun to change how they interact with their home. Finding sustainable ways to fish, grow food and maintain the forests are just a few of the ways people are changing their course to benefit the biosphere. Does this mean that more cannot be done?

No, in fact governments and individuals are producing energy in a more sustainable way. Most importantly, these official channels are changing their viewpoint about how the earth should be treated. It does not mean that the extinctions will stop overnight. But by creating the change we wish to see, we as individuals can have a greater impact on how the world and its organisms are cared for.

Conclusion

As we have seen, biology is a diverse study of life in all its forms, from the molecular to the biosphere. While we have just scratched the surface here, it is clear that the enduring theme of biology is interconnectivity.

No matter what organism we are, be it plant, animal or human, all of us are relying on each other for our continued existence. Throughout this book, we have examined biology from a variety of perspectives. We have poked into the cellular level and found out how truly busy our bodies are all the time. Throughout the chapter on DNA, it became clear that there are plenty of instructions available, even with only four letters in the DNA alphabet.

The lifecycles of animals and plants was a critical discussion, because it leads us to the issues found within the biosphere. We have continually pointed out how the biodiversity is symbiotic with not only each individual habitat, but all the creatures surrounding that particular habitat.

Even the biosphere itself has layers that interact with each other. When changes are made in one layer, those changes have consequences in other layers. Elements are moved, molecules created and destroyed, as well as multiple chemical reactions, just in a small sampling of the earth's organisms. Taken as a whole, it is amazing to see how diverse life can be, while still be so interconnected.

Take the time to explore nature and the biodiversity around your home. You will quickly find examples of the many unique processes we have discussed here, and hopefully discover many more!

Complimentary Chapters
of
Quantum
Physics

Beginner's Guide to the Most Amazing
Physics Theories

3rd Edition

Introduction

Have you ever wondered how scientists produce their explanations about light, energy and matter on molecular level? How can those same scientists measure something they cannot even see? After all, the molecular level is hardly visible to the naked eye. Quantum Physics is the study of the behavior of matter and energy on a molecular level. Think of the smallest particles we know about, such as atoms, protons, neutrons and electrons. These are the building blocks of all living things and are the smallest parts of matter and energy. When studying them, mathematics is the key to really understanding how these small parts of the world work together on a larger scale.

When using these mathematical equations, scientists find the constants within the physical laws on the molecular level and plug these constants into their equations to better understand how these physical laws act on matter and energy. Understanding how matter and energy behave allows for other real life applications to come into play.

In addition, scientists use these mathematical equations to explain what they observe in the world around them and also what they observe through various experiments. As the tools of their trade have become more precise, scientists are able to gather better information to add to their understanding of the molecular world. Today, we benefit from the work of these scientists to better understand our world and the Universe on a molecular level. As we will see, Quantum Physics is mathematics at work explaining the world around us, down to the smallest detail.

Quantum Physics has been defined by its history and the various theories this molecular study has spawn. These theories include wave particle duality and quantum tunnelling. Yet before the scientists could create these theories, there were plenty of experiments which assisted them in formulating these theories.

The experiments included a black body radiation experiment, whose observable results confounded scientists, until one researcher came up with an equation that matched the data they were observing. Other theories, such as the photoelectric effect, was the beginning of a run of experiments and hypothesis that challenged the classic wave theory. Over time, these hypotheses and experiments have built the foundation of data that is the basis for quantum physics or quantum mechanics. The two terms can be used interchangeably and we do so as this book unfolds.

The experiments discussed include the Double Slit Experiment and how it effects the Classic Wave Theory. At the same time, these experiments gave scientists the chance to observe effects that would contribute to the theories that are now part of quantum physics. Other theories highlighted within these pages include the Photoelectric Effect, the Compton Effect and even the uncertainty principle.

Throughout this book, we'll explore some of these experiments and theories, both how they came to be and then how they have grown to become critical parts of what we now know as quantum physics.

Prologue:
Exploring Quantum Physics: My Journey

One of the questions many ask is how does someone gain this much knowledge about such a complex field of study? As an individual who has a love for science and science fiction, I was compelled to dive into this unique area of study. What started as a hobby when I was kid, trying to figure out if time travel was truly possible, has grown into a lifelong pursuit to understand how all these theories meld together.

While I was good at math, the various scientific theories of Quantum Physics took work to understand and appreciate. My interest and love of the field grew over time, but was sparked when my first science teacher introduced me to Albert Einstein. For some who is new to the field or wanting a better foundation of knowledge in Quantum Physics, Albert Einstein is a great place to start.

His theories have become the base of equations to explain a variety of physics principles. Even those scientists that disagreed with him, still used his work as a base for their own. One of the most important debates was Einstein with Niels Bohr. These debates lasted over several years and included theories, explanations and challenges to the way the scientific community viewed the microscopic parts of the universe. At the same time, whatever Einstein's interest or theory, he led an entire field of study in that direction. While others can claim the title of father of Quantum Physics, Einstein is the mentoring uncle.

He continued to challenge himself and his colleagues throughout his lifetime and we are still basing a number of scientific theories on his work. So he lives on in the experiments and research being performed today. I personally find Einstein's work inspirational, because of how vigorously he defended his theories and how much he challenged those around him to think beyond the box. One of his favorite analogies was that of the lion. In Quantum Physics, it was perceived that scientists had only found the lion's tail, in the form of the equations and theories presented so far. Einstein believed that a unified field theory (the lion) did exist. His hope was that his work would assist others to find the allusive lion. As we still work to find the equation that would use all four fundamental forces and complete the puzzle, I am often reminded that we have only found the tail and maybe a foot, but someday we will have found the whole animal, thus truly understanding how all these pieces of our universe, big and small, fit together.

As we have perceived in our discussion, most of the experiments in this field of study were influenced by the observations of the scientists and researchers themselves. Thus Quantum Physics is more about explaining the microscopic world with equations to match their observations, then observations matching equations. As someone who finds this microscopic world and all the parts of it fascinating, it quickly became apparent that to understand this area of study, your math skills need to be sharpened.

After intensive study of various math, including calculus and trigonometry, one begins to understand how equations play a part in physics, but these equations are merely an attempt to describe what we, as scientists and researchers, are observing for the first time. Combining a masters in both physics and advanced math gives someone the ability to begin to break down those equations. As you understand the parts of the equations, it helps you to better understand what they ultimately represent in the world of Quantum Physics.

Much like any piece of machinery, while you can learn much from diagrams and various books, the best lessons come from the hands on aspects. For example, if you are able to take apart anything electronic, you learn much about how the components work together. It is this knowledge that can help you put the electronic back together again. Over time, your understanding of that particular machine can only grow as you pull it apart and put it back together.

With Quantum Physics, the same principles applied. By taking these equations, studying their parts and then putting them together, one's understanding continues to grow. Overtime, your explanations of the equations themselves and how they function in the real world also matures. Throughout my years of study, I attempted many of the experiments we discussed here. These experiments allow you to build your own library of observations of both the equations and how they explain real world events.

For many individuals in my area of study, physics is not easy work. After all, you are attempting to study parts of the universe so small that no one has ever really seen them. We are making observations based on events that have been recorded at various stages within the scientific community. For those who make Quantum Physics their life work, it's important to recognize that recording and measuring devices are only improving. Thus, at some point in the future, these small pieces of the universe may very well be visible to humans and not just parts of an equation. Personally, my own areas of study focus on understanding particles, especially as they relate to dimensions and potential time travel (after all these years, I am still fascinated with the time travel!).

While we wait for those amazing measuring devices of the future, our work with current tools continues. Scientific theory encourages us to study various elements one at a time, using constants to help us focus on the one piece of any particular puzzle.

This can be difficult when defining what is acting on a particular part of the equation for parts of the microscopic world. When we look at wave particle duality, for example, this conundrum becomes quite clear. After all, how can a scientist be sure that he or she has eliminated all the constants except for the particular particles or waves they are attempting to study? Our knowledge of how light moves and acts within a variety of environments continues to fascinate myself and others of our scientific community.

My personal studies have included looking at the various arguments put forth by many different scientists in my field. Quite simply, one of the joys for me of quantum physics is trying to find the better explanation for what we already know.

In many ways, quantum physics for me has been the study of possibilities. My students often laugh about how I can go on and on about various theories of Einstein's or Bohr. But they aren't laughing when given some of the problems presented by these scientists during their arguments and ask them to prove them right or wrong. These thought problems created by Einstein and others can be debated in classrooms, but also can be part of any researcher's toolbox. Call them brain stretches of sorts, but these also can help someone to achieve a different perspective on a totally different problem or difficulty with a theory.

Through my time in the field of quantum physics, it is apparent that while I have learned so much, there is still much to learn. When we look at the string theory or hidden dimensions, the limits of quantum physics become apparent. After all, we can surmise that these things exist and are true, but we simply can't prove them in a definitive fashion. But therein also lies the unique challenge of this field of study. After all, so much of quantum physics is about explaining our observations via equations.

There are also the exciting breakthroughs happening in Quantum Physics. As we discussed the M-theory, it shows how this field is changing as others bring their own theories to this community. By adjusting our viewpoint, we are able to see a new explanation for string theory. While this is just one example, there are many others. It is important to note that each of these breakthroughs then spawns years of additional experimentation and other work to gain the best understanding of what this theory means when applied in the real world.

Other areas that have helped me to grow my knowledge base include sharing with other scientists and researchers. Quantum Physics is similar to a large puzzle. Each of us is working on a different section and when we put them all together, researchers and scientists can get a complete picture of the universe. However, this particular puzzle is pretty large, so it will take a lot of scientists and researchers to complete this picture.

As a researcher and lecturer, I would encourage more young people to explore this amazing field. While this book is meant to provide a simpler view of quantum physics, my hope is that it will encourage others to join this field and help to build upon the knowledge and theories that have already been discovered.

Chapter One:
Quantum Physics: The Beginning

The Earth and the Universe, in particular matter and energy that are their building blocks, are governed according to the various laws of physics. No matter where we go or what we do, these physical laws are always in force and remain absolute. These physical processes govern how matter and energy can be transformed and its behavior in various situations where they interact with other elements or forces. Yet beyond the physical aspects of the world we can see, there is another microscopic world operating under its own set of laws, also governing the behavior of matter and energy. Scientists describe this set of laws in a group of theories known as Quantum Physics, or the study of how matter and energy behave on the atomic, nuclear and even smaller microscopic levels.

Quantum is Latin for "how much". In Quantum Physics, the quantum describes the various discrete or distinct units of energy and matter that are predicted by or observed on a microscopic level. This field of study began as scientists gained the technological tools to measure the world even more precisely, particularly the world that is not visible to the naked eye. The beginning of quantum physics, as a field of study, has been attributed to a paper written by Max Planck on the topic of blackbody radiation. Development within the field was done by various scientists, including Albert Einstein, Max Planck, Werner Heisenberg, Erwin Schrodinger, Niels Bohr and others. Let's meet Max Planck and see how his work really opened up Quantum Physics to the scientific community.

The Father of Quantum Physics

In 1874, Max Planck, a scientist who had conducted experiments in the diffusion of hydrogen through the heated medium of platinum before turning to theoretical physics, turned his attention to the ultraviolet catastrophe. This problem was based around the Raymond-Jeans formula, which was used to measure thermal radiation. This radiation is actually an electromagnetic radiation that objects produce based entirely on the object's temperature. However, the Raymond-Jeans formula was not successful at actually predicting the results of various experiments. By 1900, this formula was causing trouble for classical physics questioning the basic concepts of thermodynamics and electromagnetics, which were part of the equation. Planck reasoned the formula projected low-wavelength radiancy (otherwise known as high frequency) was significantly higher than it should be. Thus, he proposed that if one could limit the high-frequency oscillations in atoms, the corresponding radiancy of high-frequency waves would also be condensed, which would allow for consistent experimental results. This is the first example, although not the last, of scientists in Quantum Physics working to create mathematical equations that would explain what they were seeing in the natural world and through their experiments.

Planck suggested that atoms themselves can absorb or discharge energy only in specific bundles called quanta. If the energy and radiation frequency are proportional, then at higher frequencies the energy would likewise become larger. It is not possible for a standing wave to produce an energy bigger than kT. Thus, the standing wave's high-frequency radiancy is capped. By creating a cap, the problem of the ultraviolet catastrophe is resolved. While Planck may not have believed quanta was a true physical requirement, but it was a mathematical artifact that helped equations to fit the reality they were measuring.

His work provided a fundamental concept for physics, that energy exists in distinct packets that cannot be broken down any further. For example, Einstein used this concept to explain photoelectric effect in 1905, thus helping to establish the concept of the photon. However, Planck assumed that the Copenhagen interpretation was flawed and eventually, a better theory would replace his concept without the troublesome aspects of quantum theory. Instead, his work and reputation helped to cement the controversial theory of relativity as proposed by Albert Einstein. These interpretations and theories are represented as such, because while there are many different explanations of how particles and other aspects of Quantum Physics work, it can be hard to prove which explanation is the correct.

*"A new scientific truth does not triumph by convincing its opponents and making them see the light, but rather because its opponents eventually die, and a new generation grows up that is familiar with it." – Max Planck as quoted by philosopher of science Thomas Kuhn in **The Structure of Scientific Revolutions***

So what makes Quantum Physics so special within the broader scope of Physics itself? To answer that, it's important to remember that Quantum Physics uses math to explain how energy and matter behave. In other sciences, the observation of an experiment or a phenomenon does not influence the processes taking place. Yet with Quantum Physics, observation does influence the processes, because the equations are developed to explain what was observed. As the next few theories display, it's the scientists' observations that guide the overall development and the adjustments of the mathematical equations that are the brains of Quantum Physics.

Chapter Two:
Wave Particle Duality

Throughout history, science has been fascinated with light and how it behaves. Prisms, among other tools, have been used to observe and measure light. During the 1600s, Christiaan Huygens and Isaac Newton suggested opposite theories to explain the behavior of light. Huygens believed that light functioned as a wave, with various lengths. On the other hand, Newton proposed that light didn't behave as a wave, but as a particle. Newton's position in the scientific community of the time helped make his theory dominant, while Huygens dealt with issues of matching observation to his theory.

To understand how these theories differ, one has to understand how waves and particles behave. We'll use light as an example. Across the electromagnetic spectrum, light waves behave in very comparable ways. When a light wave comes across an object, it is either polarized, transmitted, absorbed, refracted, diffracted, reflected, or scattered. What happens to the light wave depends on its wavelength and the structure of the object encountered. Scientists also structure various experiments that allow them to study light by forcing it into different situations where the light is made to bounce off specific objects or bend. The data gathered becomes part of the knowledge database that others use to build their experiments and theories. So how does a wave occur?

Generally, a wave has to propagate through some type of medium. Huygens defined that medium as luminiferous aether, but today it is known simply as ether. This explanation was accepted in the scientific community, even though there was no concrete proof it existed. During the 1860s, James Clerk Maxwell quantified a set of equations (known as

Maxwell's laws or equations) to describe electromagnetic radiation along with visible light as the transmission of waves. He assumed such an ether was the medium of propagation. His predictions with this medium in mind were consistent with his experimental results. However, no such ether was ever located, but instead it remained a mystery.

Yet the scientific community could not provide an alternative that would explain the experimental results being observed without it. But as we shall see throughout these chapters, scientists continued to make breakthroughs and thus were able to craft theories that appeared to fit the phenomenon that they were observing. In addition, experiments have also built a record that current scientists are using to move this field of study ever forward.

In 1720, James Bradley completed astronomical observations in stellar aberration. He found that ether, if it existed, would have to be stationary relative the movements of Earth. Throughout the 1800s, many experiments were created to detect the ether or its movements directly, but with no success. The most famous experiment of that era was the Michelson-Morley experiment, an attempt to measure the movement of the Earth through ether. Though often called the Michelson-Morley experiment, it refers to a series of experiments first carried out by Albert Michelson in 1881. Then those experiments were carried out again with superior instruments and equipment at Case Western University in 1887, with assistance from Edward Morley, a chemist.

Light was known to travel through outer space. Scientists believed that space was a vacuum. One could create a vacuum chamber and shine a light through it. The evidence was clear that light could move through regions without air or other matter. So how could that be? Huygens' ether was the handy substance scientists used to explain how this was possible. The universe, they claimed, was filled with ether. It was this substance that gave light waves the ability to travel through space and other regions commonly lacking air or any other matter.

Michelson and Morley decided that if the ether did exist, you should be able to measure Earth's orbital rotation through it. Since ether was believed to be unmoving (static except for the vibration) while the Earth was moving quickly, it stood to reason that one could measure ether by its contact with the Earth. Therefore, researchers and scientists began to build experiments meant to capture measurements of ether in these interactions with the Earth.

Imagine for a moment holding your hand outside your window, particularly in a car. While it may not be windy, the force of your own motion (courteous of the car) makes it appear windy. Scientists believed ether should have created what would be in effect an ether wind, which would push or hinder the motion of a light wave.

To test this hypothesis, Michelson and Morley designed a scientific device, called the Michelson interferometer, which was meant to split a beam of light, then bounce it off mirrors so that the split beam moved in different directions then struck the target. The principle at work was based on the idea that if two beams traveled an equal distance, but used different paths to move through the ether, they should end up moving at different speeds. So when these beams finally hit the target screen, they would be slightly out of phase with each other, creating an observable interference pattern that could be measured. If this experiment had been successful, it would have been the first definitive proof of the existence of this ether.

The results was disappointing, however, because they found absolutely no evidence of the relative motion bias that these two scientists were hoping to observe and measure. No matter which path the split beam of light took, the light always seemed to be moving at precisely the same speed, so there was no interference to measure. Without evidence of interference, scientists found it difficult to move forward with this particular line of study. After all, with no evidence of the ether in the expected places, scientists began to think that this might not be a productive line of study.

Ether was finally abandoned with the work of Albert Einstein and his theory of wave particle duality. In 1905, Einstein published his paper explaining the photoelectric effect, in which he proposed that light travel in discrete bundles of energy (quantum). The energy contained with a photon was related the frequency of light. As a result, ether was no longer the necessary medium it had once been. But how did this explain the situations when light was observed acting as a wave, and other times when light acted as a particle?

Experiments, such as the quantum variations of the double slit experiment and the Compton Effect, seemed to confirm that light was in fact a particle. But as experiments continued and the evidence mounted, it became clear that light could act as a wave or a particle depending on the parameters of the experiment and when the observations were made. Researchers and scientists pondered how such an effect could occur. After all, they understood that matter could exist in different states, but never two or more at the same time. Yet, particles of light were doing just that. This moved the research forward, because there had to be an explanation for this occurrence. Now let's discuss how the wave particle duality translated into from light to matter.

Wave Particle Duality in Matter

Scientists wondered if matter would also show such duality. The de Broglie hypothesis was an extension of Einstein's explanations of matter's wavelength in relation to its momentum. For de Broglie, Einstein's relationship of wavelength to momentum seemed able to determine the wavelength of any matter. His reasoning for choosing momentum over energy is based on the various energy types available to use in the equation, such as total, kinetic or total relativistic energy. For photons, it wouldn't matter because all energy is the same in that instance. But matter is different and so momentum was this 1929 Noble Prize winner's choice.

Just like light, it seemed that matter also exhibited both wave and particle properties under precise circumstances. Obviously, massive objects would exhibit very small wavelengths. But for small objects, it is possible to observe the wavelength, as noted in the double slit experiment with electrons. So now the same behavior was being observed in several different settings. Researchers and scientists recorded the data, and then began to study it to determine why these effects were occurring, attempting to come up with a theory and equation to fit the observations.

But what does it matter if light or matter acts as a wave and a particle?

Significance of Wave Particle Duality

The major significance of this theory is that all behavior of light and matter can now be explained through an equation that denotes wave function, generally found in the form of the Schrodinger equation. As a result, describing reality in the form of waves is the heart of quantum mechanics, the mathematical brain of quantum physics.

The most common interpretation of this theory is that the wave function simply represents the probability of locating a given particle at a given point. These probability equations can exhibit supplementary wave-like properties, creating a concluding probabilistic wave function exhibiting these properties also. In other words, the probability of a particle being present in any location is a wave, but the actual physical appearance of that particle isn't a wave at all. Instead, it is a particle but not until the moment that it is measured in that particular place or space.

The complicated math can result in fairly accurate predictions, the physical meaning of these equations are much harder to grasp. Explaining what the wave particle duality really means continues to be a key point of debate. These debates have created multiple interpretations to explain this particular duality, but at the same time these interpretations are bound by unambiguous wave equations, and are required to explain the same experimental observations. No easy task, as science continues to dig into what this theory means to the real world.

As studies continued in the realm of quantum physics or mechanics, evidence of other types of behavior by electrons and atoms began to mount. Scientists worked to discover what the cause of these effects was. As a result, the many areas of study in quantum physics began to develop. In the next few chapters, we will be discussing some of these areas of study but also some of the more famous experiments in the field of Quantum Physics. The first one we will discuss is quantum tunneling.

Chapter 3:
Quantum Tunnelling

As a result of the wave-particle duality, it can appear that particles pass through walls. The phenomenon has been well documented and the process is understood within the rules of quantum mechanics.

Quantum tunnelling (or tunneling) is the quantum-mechanical outcome of transitioning into a previously-forbidden energy state. Consider rolling a ball up a hill. If the ball does not have the proper amount of velocity, then it will not roll over the hill. This makes sense to many of us who have read the tales of Greek mythology, particularly Sisyphus and his endless quest to roll his boulder up the hill. While he was not successful, many scientists believe that there might be another way for the ball to get to the other side of the hill. It just isn't a traditional one. Why? Because quantum mechanics offers another solution.

In quantum mechanics, objects do not behave like classic objects, but instead exhibit a wave like behavior (as we discussed in Chapter 2). In thinking of a quantum particle, since is it both a wave and a particle, the particle can in theory extend through the hill because of its wave like qualities. Various probability equations can predict the probability of the particle's location and it has the possibility of being detected on the other side of the hill. As a result, it appears to have tunneled through the hill, thus the name quantum tunnelling or tunneling.

Scientists measured electrons escaping that should not be energetic enough to make a break for it. In the normal world around us, this would be similar to a child jumping into the air, but instead leaping over the whole house (gravity not withstanding). The child should not be expected to achieve that feat based on their specific skill set. Scientists also were puzzled because they were measuring something that didn't seem achievable by the electrons. So how were they making the leap? It is possible because of the quantum tunneling. This unique way of moving takes advantage of the various natures or states of matter.

Quantum tunneling is possible because of the wave-like nature of matter. As confusing as it may seem, in the world of quantum physics, particles can perform actions that are similar to waves of water rather than billiard balls. To put it simply, an electron doesn't exist in one place at one specific time with a defined amount of energy, but instead exists within a wave of probabilities. As a result, the particle acts more like a wave and appears to flow in a wave like fashion. Probability predicts the various points of a wave or where a particle will be at any given point in time.

The physicist Manfred Lein stated that electrons can be designated by wave functions. These functions can extend from the inside to the outside of an atom, demonstrating that a portion of an electron is always on the atom's outside.

In one recent experiment, researchers used a laser light to subdue the energy barrier that would typically trap an electron inside a helium atom. This laser reduced the overall strength of the barrier so that an electron wouldn't have the energy required to escape the atom. Instead, the atom could try to cheat and similar to a mole tunnel its way through. The researchers found that the electron tunneled through in a very short window of time. They are currently trying to trace the cycle of the electron. By doing so, they hope to determine the exact moment the electron officially left the energy barrier. So how will they measure something so infinitely small?

To measure this, these physicists looked for the photon of light produced when an electron rejoins the atom after making its escape through the tunnel. In some instances, scientists have used a laser to keep the electron away, thus preventing it from recombining with the atom. By doing so, they are able to observe the electron's tunneling and take the appropriate measurements. The forced separation also gives the scientists the ability to note how the electron reacts when forced to remain separated from its original atom.

While this is the first time scientists pinpointed when an electron tunneled through an atom, it won't be the last. Today, technology is providing scientists with ever more accurate tools to help them measure and understand the molecular world. Previously, theoretical calculations could predicted the timing of quantum tunneling, but the process had not been directly measured and with such accuracy.

The findings could help scientists understand other speedy courses that count on quantum tunneling, which are often observed within nature itself. These experiments are just part of a larger attempt to understand how the Earth and the Universe function within the limits of physical laws. But Quantum Physics also has its central principles that help to define the world around us. We will discuss a few in Chapter 4.

Chapter 4:
Quantum Entanglements, Quantum Optics and Electrodynamics (QED)

Our parents often told us to watch who we associated with, because it would reflect either poorly or positively on us. One of the bedrock principles of Quantum Physics is Quantum entanglement, though it is also highly misunderstood. In short, Quantum entanglement means multiple particles are linked together in a way that means the measurement of one particle's quantum state controls the possible quantum states of the other particles within the linked group. As such, these particles act on one another, much as the friends of our younger selves did. Let's look at this principle a little closer to understand what science has observed.

The Classic Quantum Entanglement Example

The classic example of quantum entanglement is called the EPR Paradox. The EPR Paradox (or the *Einstein-Podolsky-Rosen Paradox*) is a thought experiment intended to exhibit the inherent paradox in the early formulations of quantum theory. This thought experiment is among the best-known examples of Quantum entanglement. The paradox involves two particles that are entangled with each other according to quantum mechanics. Under the Copenhagen interpretation of quantum mechanics, each particle is independently in an uncertain state until it is measured, at which point the particle's state becomes certain.

At that exact same moment, the other particle's state also becomes certain. The reason that this is classified as a paradox is based on the fact that it appears the two particles must have communicated at speeds greater than the speed of light, a conflict with Einstein's theory of relativity. This paradox was at the heart of a debate between Albert Einstein and Niels Bohr.

In the more popular Bohm formulation of the EPR Paradox, an unstable spin 0 particle decays into two different particles, Particle A and Particle B, both heading in opposite directions. Because the initial particle had spin 0, the sum of the two new particle spins must equal zero. If Particle A has spin +1/2, then Particle B must have spin -1/2 and vice versa in order to equal zero. According to the Copenhagen interpretation of quantum mechanics, until a measurement is made, neither particle would have a definite state. Both particles are in a superposition of possible states, with an equal probability of having positive or negative spin.

There are two key points within this paradox that make it troubling to scientists.

1. Quantum physics explanations state that until the moment of the measurement, the particles do not have a definite quantum spin, but instead are in a superposition of possible states.

2. Upon measuring the spin of Particle A, we know for sure the value we'll get from measuring the spin of Particle B.

Another words, whatever Particle A's quantum spin is set by a measurement, then Particle B must somehow instantly know what the spin is that it is supposed to take on. As Einstein pointed out, this is a clear violation of his theory of relativity.

Niels Bohr and others defended the standard Copenhagen interpretation of quantum mechanics, as supported by experimental evidence. The explanation is that the wave function which describes the superposition of possible quantum states exists at all points simultaneously. The spin of Particle A and the spin of Particle B are not independent quantities, but are represented by the same term within the equations. The instant the measurement on Particle A is made, the entire wave function collapses into a single state. Therefore, no communication is occurring at the speed of light.

This relationship means that the two particles are entangled. When you measure the spin of Particle A, that measurement has an impact on the possible results you could get when measuring the spin of Particle B. This has been verified by Bell's Theorem.

A fundamental property of quantum theory is that prior to the act of measurement, the particle does not have a definite state, but is in a superposition of all possible states. Imagine for a moment a cat in box with limited oxygen. Because the cat is unobserved, the cat is both dead and alive, since there is no way to definitively say what the cat's state is. Yet, upon opening the box, the cat's state is immediately defined, just as when a particle is measured and its position is clearly defined.

Though this interpretation does mean that the quantum state of every particle in the universe affects the wave function of every other particle, it does so in only mathematically. There is really no sort of experiment which could ever truly discover the effect in one place showing up in another. We have discussed light and waves throughout these chapters, but now it's time to look at the specialized study of light or photons and their interaction with matter.

Quantum Optics

Quantum optics is a field of Quantum Physics dealing specifically with the interaction of photons with matter. The theory is that light moves in discrete bundles or photons as represented by Max Planck's ultraviolet catastrophe paper (see Chapter 1). As Quantum Physics developed through the early part of the 20th century by understanding how photons and matter interacted and were inter-related. This was viewed, however, as primarily as a study of matter, not necessarily light.

In 1953, the maser was developed that emitted coherent microwaves. During 1960, the laser made its appearance, known for emitting coherent light. Using these tools, with a focus on light, Quantum Optics was used to describe this specialized field of study.

The findings of Quantum optics support the view of electromagnetic radiation as traveling in both forms, a wave and a particle, as what we have learned as wave particle duality. By using the findings from quantum electrodynamics (QED), it is possible to define quantum optics in the form of the creation and annihilation of photons.

This approach allows the use of certain statistical approaches to analyze the behavior of light, although whether it represents what is physically taking place is a matter of some debate.

Lasers and masers are the most obvious application of quantum optics. Light emitted from these devices is in a coherent state, which means the light resembles a classical sinusoidal wave. In this coherent state, the quantum mechanical wave function and its uncertainty is distributed equally. Laser light is highly ordered, and generally limited to essentially the same energy state, and by default, the same frequency and wave length.

Quantum Electrodynamics (QED)

Quantum electrodynamics (QED) is the theory of the interactions of charged particles with an electromagnetic field. These interactions are described mathematically, not just interactions of light with matter but also those of charged particles with one another. QED is a relativistic theory, because Einstein's theory of special relativity is built into each of the equations. Because the behavior of atoms and molecules is principally electromagnetic in nature, all of atomic physics are considered a test laboratory for the theory. Some of the most precise QED tests are experiments dealing with the properties of subatomic particles known as muons. The magnetic moment of this particle type has been shown to agree with the theory to nine significant digits. Agreement of such high accuracy makes QED one of the most successful physical theories so far devised.

The interaction of two charged particles occurs as part of a series of processes building into increasing complexity. In the simplest, only one virtual photon is involved and each process adds virtual photons. The processes correspond to all the possible ways that the particles can interact through the exchange of virtual photons. Each of these can be represented graphically by means of the so-called Feynman diagrams. Besides furnishing an intuitive picture of the process, this type of diagram prescribes precisely how to calculate the variable involved. Each subatomic process becomes computationally more difficult, and there are an infinite number of processes. The QED theory states that the more complex the process, the smaller the probability of its occurrence.

QED is often called a perturbation theory due to the smallness of the fine-structure constant and the resultant decreasing size of the higher-order contributions. This relative simplicity and the success of QED has made it a model within quantum field theories. Additionally, the picture of electromagnetic interactions as the exchange of virtual particles has carried over to the theories of the other fundamental interactions of matter, the strong force, the weak force, and the gravitational force. But with all these theories floating around, how does one fit them together. The Unified Field Theory is an attempt to create a single theoretical framework, as we'll learn about in Chapter 5.

Chapter 5:
Unified Field Theory

So what is Unified Field Theory? Albert Einstein first coined the term to describe any attempts to unify the fundamental forces of physics, particularly between elementary particles into a single theoretical framework. Einstein himself searched for such a Unified Field Theory, but was not successful. So what brought this about?

In the past, seemingly different interaction fields or forces appeared to have been unified together. For example, James Clerk Maxwell successfully unified electricity and magnetism into electromagnetism in the 1800s. In the 1940s, Quantum electrodynamics translated his electromagnetism into the terms and mathematical equations of Quantum mechanics. During the following decades, physicists successfully unified strong nuclear interaction and weak nuclear interactions, along with Quantum electrodynamics to create the Standard Model of Quantum Physics.

The current problem with a fully unified field theory is in finding a way to incorporate gravity, which is best explained by Einstein's theory of general relativity, with the Standard Model that describes the quantum mechanical nature of other three fundamental interactions. The curvature of space time, fundamental to general relativity, leads to difficulties in the quantum physics representations of the Standard Model.

Some specific theories that attempt to unify quantum physics with general relativity include:

1. Quantum Gravity - Generally is posed that a theoretical entity or a graviton, which is a virtual particle that

mediates the gravitational force. This is what distinguishes quantum gravity from certain other unified field theories. In fairness, some theories typically classified as quantum gravity don't necessary require a graviton.

2. String Theory – This uses a model of one-dimensional strings in place of the particles typically used in Quantum Physics. These strings vibrate at specific resonant frequencies. The formulas resulting from string theory predict more than four dimensions, but the dimensions are curled up within the Planck length.

3. Loop Quantum Gravity – This theory seeks to express the modern theory of gravity in a quantized format. The approach involves viewing space time as broken into discrete chunks. It is viewed by many as the well-developed alternative to quantum gravity outside of string theory.

4. Theory of Everything – This theory is a hypothetical single, all-encompassing, coherent theoretical framework of physics explaining and linking together all physical aspects of the universe.

5. Supersymmetry – A theory of particle physics, is a proposed type of space time symmetry relating two basic classes of elementary particles. The first are bosons, which have an integer valued spin. The other is fermions, which have a half-integer spin. A particle from each group associates with each other, creating a superpartner, with a spin differing by a half integer. Perfectly unbroken supersymmetry, in theory, means that each pair of superpartners shares the same mass

and internal quantum numbers, in addition to their spin.

As these theories show, the idea of one unifying theory has been difficult to prove and hard to identify. Unified field theory is highly theoretical, and to date there is no absolute evidence that it is possible to unify gravity with all the other forces. Historically, other forces have been combined, and many physicists are willing to devote their lives, careers, and reputations attempting to show that gravity can also be expressed quantum mechanically. The magnitudes of such a discovery, of course, cannot be fully identified until a viable theory is proven by experimental evidence.

Chapter 6:
Black Body Radiation

The wave theory of light was the dominant light theory in the 1800s. This theory was captured by Maxwell's equations and surpassed Newton's corpuscular theory. However the theory was challenged by how it explained thermal radiation, which is an electromagnetic radiation given off by objects based on their temperature. So how could someone test or detect thermal radiation?

Scientists can test for thermal radiation by setting up an apparatus to detect radiation from an object based on specific temperature, represented by T_1. Warm bodies give off radiation in all directions, so in order to be able to measure it effectively, shielding must be used so the radiation is examined in the form of a narrow beam.

In order to create this narrow beam, a scientist use a dispersive medium, such as a prism, placed between the body or object emitting the radiation and the radiation detector. This allows the radiation wavelengths to be dispersed at an angle. Then the detector measures a specific range or angle, essentially the narrow beam. This beam is considered a representation of the total intensity of the electromagnetic radiation across all the wavelengths.

So let's define a few key points. One thing to note is that the intensity per unit of a wavelength interval is referred to as radiancy. Calculus notation helps us to reduce various values to zero and create the following equation: $dI = R(\lambda)\,d\lambda$. Using the prism, a scientist can detect dI, or the total intensity over all wavelengths, so one can define radiancy for any wavelength by working backwards through the equation. Now let's look at how we can build a database of sorts for wavelength versus radiancy curves.

Scientists typically perform an experiment over and over again, building up a store of data that creates various ranges. When working with these ranges, one can begin to build a better understanding of how much radiation will occur from a specific object, but also how intense it will be at any given temperature.

For instances, one can glean that as the total intensity radiated increases as we increase or decrease the temperature. But when we look at the wavelength with the maximum radiancy, we find that the inverse occurs, that is with that specific wavelength, the intensity will go down as the temperature increases. Thus, as the temperature go up, wavelengths can change their individual radiation intensity, but the overall radiation intensity will continue to increase with the temperature.

So if the temperature is going to down, then the maximum intensity of an individual wavelength will go up, but the overall or total intensity of the object will go down, corresponding with the temperature.

Again, we go back to how to measure something when light reflects off so many things. How do you create the angle, making sure that you are accurately measuring your narrow beam?

A simple way to do this is to stop looking at the light and look at the object that doesn't reflect it. Light does reflect off of objects, but scientists will perform this experiment observing a blackbody, or an object that doesn't reflect any light at all.

Otherwise, the experiment runs into a difficulty defining what is being tested.

Performing this experiment requires a box, preferably metal, with a tiny hole. If or when light hits the hole, it enters the box but it won't bounce back out. As a result, the hole, not the box, is the blackbody of the experiment. Any radiation detected outside of the hole is a radiation sample of the amount of radiation in the box. Scientists analyze this information to understand what's happening within the box.

The first thing to be noted is that the metal box is being used to stop the electric field at each wall of the box, creating a node of electromagnetic energy at each of the walls. Thus standing electromagnetic waves are contained within the box.

Second, the number of standing waves with their various wavelengths within a defined range including an equation that takes into account the volume of the box. By analyzing the standing waves and then following this equation, it can be expanding into three dimensions.

Third, classical thermodynamics contributes a basic truth: the radiation in the box is in a thermal equilibrium with the walls of the box at a certain temperature. The radiation within the box is absorbed and reemitted from the walls constantly, creating oscillating within the radiation's frequency. The thermal kinetic energy of these oscillating atoms are simple harmonic oscillators, so the mean kinetic energy equals the mean potential energy. As a result, each wave contributes to the total energy of the radiation within the box.

Fourth, energy density is related to the radiance. Energy density is defined as the energy per unit volume within the relationship. The measurement of this is determined by the amount of radiation passing through a component of surface area with a cavity.

Classic physics as represented by the Rayleigh-Jeans formula failed to predict the actual results of these experiments, primarily due to the fact that classic physics failed to account for shorter wave lengths. At longer wavelengths, the Rayleigh-Jeans formula more closely matched the observed data. This failure was referred to as the ultraviolet catastrophe. In early 1900, this was a big issue, because it called into question such basic concepts as thermodynamics and electromagnetics as part of that equation.

Historically, this is where quantum physics came into play. Simply, Max Planck used quanta to create what would be defined as discrete bundles of energy. Thus the quanta would be proportional to the radiation frequency. With this theory, no standing wave could have more energy than kT, then high radiation frequency would be capped, solving the ultraviolet catastrophe. In the end, frequency describes the energy of each quanta, where a proportional constant.

While this resulted in an equation that fit the data of the experiments perfectly, but it wasn't as attractive as the Rayleigh-Jeans formula. This formula became the starting point of quantum physics as we know it today. Einstein even demarcated it as a central principal of the electromagnetic field, while Planck had originally used it just to solve the issue of one experiment. While it took scientists a while to warm up to what is now known as Planck's Constant, it is now considered a critical part of the quantum physics or quantum mechanics.

This was just one part of the large array of experiments that define quantum physics. Another early experiment in concert with wave particle duality, a challenge that was known as the photoelectric effect.

Chapter 7:
Photoelectric Effect

This significant challenge appeared in the study of optics during the 1800s. The *photoelectric effect* tested the *classical wave theory* of light, which was predominant at that point in time. In coming up with the solution to this dilemma, Einstein would gain a reputation in the physics community, eventually earning the Noble Prize in 1921.

The Photoelectric Effect was first observed in 1839, it wasn't documented until 1887 in a paper by Heinrich Hertz. Basically a light source is incident upon a metallic surface, the surface emits electrons called photoelectrons.

In order to observe this effect, scientists would create a vacuum chamber with photoconductive metal at one end, plus a collector at the other end. By shining a light on the metal, electrons are released that move through the vacuum to the collector. As a result, a current is created in the wires connecting the two ends. This current is measured with an ammeter. When administering a negative voltage potential to the collector, more energy is expended for the electrons to complete their journey, thus initiating a current. When no electrons find their way to the collector, this is called the *stopping potential* V_s, which can be used when defining the maximum kinetic energy of the electrons themselves.

Note that not all electrons have this energy, but will have a range of energies based upon the experimental metal's properties. The equation created to calculate the maximum kinetic energy of the particles bumped free of the metal surface at a maximum speed.

Classic Wave Theory explained as energy of electromagnetic radiation being carried within the wave itself. As the wave collided with the metallic surface, the electrons absorb the wave's energy until it exceeds the binding energy, thus releasing that electron from its metal surface.

This classic explanation includes three fundamental predications:

1. The resulting maximum kinetic energy should have a proportional connection with the strength of the radiation involved.

2. This effect will occur with any light, regardless of wavelength or frequency.

3. A delay will be witnessed of seconds between the radiation's contact with metal and its first release of photoelectrons.

Yet the experimental results directly contrasted with these three predications.

1. Maximum kinetic energy of the photoelectrons wasn't effected by the intensity of the light source.

2. The photoelectron effect was not observed below a certain frequency.

3. The delay was not observed between the radiation's metallic contact and the emission of the first photoelectrons.

Since these are the exact opposite of what was predicted by the wave theory and completely counter-intuitive to what the scientists believed would occur. Einstein would publish a paper in 1905 that built on Max Planck's black body radiation theory to explain the photoelectric effect and the contradictions scientists were observing. What he proposed was that radiation energy is not distributed equally over the wave front, but is localized into smaller bundles called photons.

The photon's energy was associated with its frequency, along with its proportionality constant and the speed of light. The proportional constant could also be defined using the wave's length.

According to Einstein, a photoelectron was released from an interaction with a single photon, rather than interacting with the whole wave. The energy transfers instantaneously to a single electron, knocking that electron free if the energy is high enough to break away from the metal's work function. If the energy or frequency is too low, there won't be any electrons released. With excess energy, beyond what is available in the photon, this excess energy will be converted into the photon's kinetic energy.

So what Einstein put forward is that maximum kinetic energy is completely independent of light intensity. He was so confident that he didn't even add the intensity of light into his equation.

When researchers shine twice as much light, they get twice as many photons, but the maximum kinetic energy itself doesn't change unless the energy of the light, instead of its intensity changes. So when does the maximum kinetic energy occur? It results when the least-tightly bound electrons break free. As for the more tightly bound photons, when they are knocked free there is a result of kinetic energy equal to zero. The result was equations that indicate why the low-frequency light couldn't free any electrons, thus producing no photoelectrons.

Since Einstein, other experimentation has been carried out by Robert Millikan, not only confirmed Einstein's theory, but also won Millikan a Nobel Prize in 1923. This experiment and the resulting data helped to crush the classic wave theory, as it was shown that light behaved as both a wave and a particle, commonly known now as the wave particle duality.

But other scientists were studying light and proving the wave theory with their experiments. One such experiment was the Young Double Slit Experiment.

Made in the USA
Middletown, DE
18 August 2017